THE FULLNESS OF TIME

THE FULLNESS OF TIME

Marking the Day by Birdsong, Blooms, Shadows and Stars

CATHY HAYNES

BLOOMSBURY PUBLISHING
LONDON · OXFORD · NEW YORK · NEW DELHI · SYDNEY

BLOOMSBURY PUBLISHING
Bloomsbury Publishing Plc
50 Bedford Square, London, WC1B 3DP, UK
Bloomsbury Publishing Ireland Limited,
29 Earlsfort Terrace, Dublin 2, D02 AY28, Ireland

BLOOMSBURY, BLOOMSBURY PUBLISHING and the Diana logo
are trademarks of Bloomsbury Publishing Plc

First published in Great Britain 2026

Copyright © Cathy Haynes, 2026

Cathy Haynes is identified as the author of this work in accordance with the Copyright,
Designs and Patents Act 1988

All rights reserved. No part of this publication may be: i) reproduced or transmitted in
any form, electronic or mechanical, including photocopying, recording or by means of
any information storage or retrieval system without prior permission in writing from the
publishers; or ii) used or reproduced in any way for the training, development or operation
of artificial intelligence (AI) technologies, including generative AI technologies. The rights
holders expressly reserve this publication from the text and data mining exception as per
Article 4(3) of the Digital Single Market Directive (EU) 2019/790

All images are courtesy of the author unless otherwise stated. p.59: Waulking Cloth, Eriskay, photograph by Werner Kissling, Ref: BVIII/3/F6/5415, Neg: F/102/30, The School of Scottish Studies Archives, The University of Edinburgh; p.65: Margaret Fay Shaw's photograph of Peigi MacRae milking Dora the cow on the island of South Uist in the Outer Hebrides, c. 1930, National Scottish Trust; p.84: St Gregory's Minster dial, in Kirkdale, North Yorkshire, courtesy of Rob Ainsley; p. 92: sketch after Frank Poller's draughts in *Medieval scratch dials on Vale of White Horse churches*, 3rd issue, Frank Poller, 1996, by Carmen Balit; p.101: Saxon pocket sundial: Science and Society Picture Library/Getty Images; p.104: Mario Arnaldi's hand dial illustration after a photograph by Marco Rech in Gabriele Vanin, *Le meridiane bellunesi, quaderno n.9, Comunita Montana Feltrina – Centro per la documentazione della cultura popolare*, Libreria Pilotto Editrice, Feltre, 1991, p.23; p. 129: diagram reproduced in the likeness of fig. 2.7 in Duncan Garrow and Neil Wilkin, *The World of Stonehenge*, British Museum, London, 2022, p. 81, by Carmen Balit; pp. 148 and 149: diagrams in the likeness of those represented in Marcel Minnaert's *Light and Color in the Outdoors*, Springer, 1993, p. 294, figs. 169 and 170, by Carmen Balit; p.181: *Le Grant Routtier*: via CC0 1.0 Universal; diagrams on pp. 87, 110, 142 by Carmen Balit; star maps on pp. 175, 177, 188, 195 by John A. Paice.

Bloomsbury Publishing Plc does not have any control over, or responsibility for, any
third-party websites referred to in this book. All internet addresses given in this book were
correct at the time of going to press. The author and publisher regret any inconvenience
caused if addresses have changed or sites have ceased to exist, but can accept no
responsibility for any such changes

A catalogue record for this book is available from the British Library

ISBN: HB: 978-1-5266-6513-3; EBOOK: 978-1-5266-6510-2

2 4 6 8 10 9 7 5 3 1

Typeset by Six Red Marbles India
Printed and bound in Great Britain by Clays Ltd, Elcograf S.p.A

To find out more about our authors and books visit www.bloomsbury.com
and sign up for our newsletters

For Rosie Hope

And in memory of my parents

Contents

Introduction	1
1 Dawn Song and Bat Flit: Heeding the habits of fellow creatures	11
2 Day's-Eyes and Turnsoles: Following the hours of flowers	31
3 Waulking Songs and Furlong-Ways: Beating time by the human voice and body	57
4 Scratch Dials and Stick Dials: Dividing the day by the flow of shadows	82
5 Daymarks: Tracing time by the light over the landscape	113
6 The Gloaming and the Dimpse: Sensing time by the colours and qualities of twilight	139
7 Star Clocks: Marking time by the motion of the night sky	168
Epilogue	199

Acknowledgements	201
A note on the text	209
Notes	211
Index	247

Introduction

We used to tell time by shadows shrinking or the midday glow over a mountaintop. We called a phase of darkness 'cockcrow' and named lively flowers for when they open and close. When working, we may have synchronised a task by singing. We'd notice the quality of the light changing at dusk, and we might mark time at night by the motion of the stars.

Now we depend heavily on complex machines and think of the hours as identical abstract units. In the urbanised West today, so much of our world would not function without super-precise clocks. But what riches might we gain from exploring the forgotten art of sensing time by events in the world around us?

This book is an adventure in discovering how past generations told the time of day by signs from the living world or used their own bodies and voices to judge spans of time. In our increasingly clock-bound and screen-immersed existence, it's an invitation to notice subtle changes in the living world, even in the tarmac-and-glass heart of the twenty-first-century city. Over the coming chapters we will sense time through the day from the dawn chorus, flowers folding up at noon, singers beating the rhythm of a task, the slow slide of shadows and sunbeams, the air turning golden, our planet spinning against the stars. While our world may still tick most loudly to the beat of the clock, by becoming more attentive to signs like these, I hope we might deepen our awareness of the fullness of time.

I've lived my whole adult life in the city – my natural habitat. I've spent my days working indoors for art organisations in London (and, briefly, below the streets as a curator for Art on the Underground). But I grew up on a traditional mixed farm in the English Midlands among people intensely aware of the changes in the living world through the day and seasons.

My days as a child were shaped by the robin singing in the darkness of early morning, my dad calling his sheepdog when he went out shepherding, the spring blooms in Mum's garden. The ever-changing events in the field and flowerbed gave broad shape to the days and seasons. Those fluid rhythms, of course, don't provide the solid pattern of the clock and calendar, which coordinate many activities on a modern farm. But there is only so much you can schedule ahead of time. The weather determines the golden opportunity, for example, to cut the barley.

When my father was in his eighties, I happened to ask if he ever glanced at the evening sky for a sense of tomorrow's weather. He replied with a grin that he preferred the Met Office forecast. Dad kept a sophisticated barometer by his chair and, in recent years, had confirmed the readiness of crops with a precision moisture meter. Yet a week or two before he brought out the meter, we'd watch him get a feeling

INTRODUCTION

for the time to harvest by walking into the barley or wheat field and selecting a sample. He'd flake the ear of corn to shift the chaff, then test the grain between his teeth. He'd give us some grains, too, to test if they were still soft and milky or dry and nutty. Dad combined high-tech tools, it seems to me, with older ways of sensing the right time to act. This is the kind of time you plan for but can't entirely predict.

What's more, Mum and Dad both had a rich store of weather lore for every part of the year, though they didn't rely on it. A few years ago, I asked them to write down those sayings whenever they thought of them. Among the long list they made, my favourites include: 'When you can sit on the ground with a bare behind, it's fit for spring corn planting', 'Rain in June puts all in tune', and (as it turns out, the flawed advice) in November, 'If ice supports a duck, all winter sludge and muck.' There are similar proverbs around the country. But some of the signs of what's coming over the horizon are very local, and well understood by people with intimate knowledge of the patterns of a particular place. Dad remembered a herdsman saying to him as a child, 'It's going to rain, m'lad! Look at the cloud in Kirby Hole.'[1]

Mum and Dad were, I think, part of a transitional generation. They followed the newspaper weather forecast. They wore wristwatches and read the time from clocks, phones, radio, TV, computers and church bells. But they could remember when clocks were not so ubiquitous and when you didn't always carry a watch. And their feeling for time and season was continually shaped by the available light, the state of leaf and bud, the colour of the corn, the warmth and dampness of the soil, the patterns of bird flight and song, the rhythms of cattle and sheep.

As a young person, I left the countryside and moved to the city. Many of the skills of observation I learned from my family in childhood I applied to looking at artworks. And somewhere along the line I stopped giving such deep attention to my everyday surroundings.

After more than a decade of living in London – a place I love – I began to feel more intensely the absence of the rhythms of life I'd grown up with. In May 2008, in a break between jobs, my partner Rosie and I took off in a camper van to the North of England. We brought very few books apart from the wildlife guides, full of fine hand-drawn illustrations, that Mum had given me in childhood. They were well-worn and much treasured, but in adulthood I'd only sporadically taken them off the shelf.

Rosie and I turned off our phones for days at a stretch (a more socially acceptable thing to do in 2008). And for the first time in adulthood, we did nothing but focus on the unfamiliar world around us for a whole summer. Without bright light to banish the night, we fell into rhythm with the daylight, as Rosie said, and drifted to sleep at dusk. Without bricks and tiles to muffle sound, we felt like the Mole in Kenneth Grahame's *The Wind in the Willows* when he emerges from the sensory seclusion of his underground home. The birdsong hits the Mole 'almost like a shout', and he leaps for 'the joy of living'. We didn't exactly leap, but we did feel the joy of being alive in a more-than-human world. We saw the stars, were rained on nearly every day, thrilled at puffins plunging into the sea, admired the light and shadow in the folds of the hills, and carried my pocketbook of flowers to botanise on the moors.

Much later, during the Covid-19 lockdowns – when our usual routines collapsed into an eerie, fearful limbo – many

INTRODUCTION

of us became newly aware of and enthralled by the exuberant worlds of other species. We began to notice what was happening around us in plain sight. And for Rosie and me, in some ways that sudden falling of the veil felt familiar from our experience back in 2008.

Our van trip to the North wasn't a return to farming life, whose seasonal rhythms are different from most urban jobs. We were holidaying while the haymakers around us were toiling. And after a few weeks, when our own work started to pick up again, we went home to London. But that sense of being part of a much larger world never really left us, and started to influence our life.

For ordinary people, navigating your way through the times of day and night may once have required understanding a little practical astronomy, staying sensitive to your body's orientation in space, knowing the habits of birds, having a keen eye for the quality of the light. To grasp something of that variety of knowledge and skill, I've roamed beyond my field and sought the help of many generous specialists. What I've written here draws insights from expertise as diverse as archaeology, architecture, astronomy, botany, literature, mythology, ornithology, social history, shepherding, sundial-making, work songs and zoology.

There has been a long-standing perception that the arrival of the clock quite quickly produced a homogenous shift in how people thought about and measured time. In recent years, historians have begun revealing a more varied and nuanced picture of attitudes and practices on the ground. Rather than seeking to exclude mechanical timekeeping, I'm following those who've 'decentred' the clock. I've moved complex instruments to the edge of the frame and concentrated on investigating how people in (mostly northwestern) Europe told the time of day by their own everyday methods using simple tools or their own senses, even in recent centuries. These old ways are too easily overlooked and forgotten but may once have seemed as natural as glancing at a screen or dial feels to us now.

My search for ways of marking time by simple, sensory means has often focused on the UK, partly because this is where I've always lived. It's the place I know best and whose traditions I can most easily access. And I was intrigued by the challenge of looking for alternatives in a culture where scientific timekeeping has been so dominant for so long. This was, after all, a nation which historically imposed its clock-time standards on its colonies, and which exported Greenwich Mean Time to the globe.[2] I sought to test my expectation that it would be too hard to find traces here of other ways of marking time. And I'm glad I did.

Another crucial reason for most often staying close to home is that – in my indoor, screen-focused life – I've become less and less aware of the patterns that would have shaped the rhythm of the day for past generations. That is, the voice of a bird, the slant of light on a wall, the glint of the evening star. I wanted to investigate those signs in my own surroundings – and as much as possible in the flow of ordinary urban working life.

In the pages that follow, we listen to the dawn chorus and hunt for bats emerging at dusk under a flightpath. We discover surprising events among the plants in the pavement cracks and speak to a scientist who's tried to grow a clock from living flowers. We listen to the rhythms of the songs people sang in the Hebrides while finishing tweed cloth or milling grain by hand. We learn how early twentieth-century English shepherds made sundials in the turf. We seek medieval time-markers in the snow with a leading Icelandic historian. We discover how a 1970s English town came to be oriented to midsummer sunrise. We sense our way through the internal phases of twilight with an astronomer who's been closely observing the transition from daylight to darkness for decades. And we raise our gaze to the stars above the streetlights in inner London.

We're not the first generation to feel that our dependence on the clock places a baffle on the senses, narrowing perception of other rhythms of time on the land and in the sky. Early in the nineteenth century, for example, William Hazlitt composed an essay in praise of the sundial, where he writes that those without 'artificial means' to tell the time are 'in general the most acute in discerning its immediate signs'.[3] In Thomas Hardy's novel *A Pair of Blue Eyes* (1872–3), Mrs Swancourt appears to echo Hazlitt when she remarked, 'how truly people who have no clocks will tell the time of day'. This is such an abiding concern in Hardy's work that it's felt as if, almost wherever I've turned, he's been ready and waiting with a vivid portrait of people telling time by other means: the Sun moving over the trees, the routines of birds, the regular sound of the cows being called at milking time, the 'tints and traits' in the landscape, the 'pinking-in' (dimming) of the day.[4] In the final chapter of this book we take lessons on telling time by the night sky from his novel *Far from the Madding Crowd* (1874).

Each chapter is chiefly focused on times of day and night, or times shorter than a day. And there is a seasonal dimension, too, because signs from the world around us change prominence as the year turns. In winter in my part of Europe, birds are quieter, flowers fade, shadows are scant and twilight falls early, but on clear evenings the brightest stars shine even in the inner city. As the year warms up, flowers bloom, the dawn chorus grows and bats come out of hibernation. On summer days, sunbeams and shadows are, naturally, most abundant. That ebb and flow of rhythms is no surprise. But in the course of this adventure, I've found how easy it is to miss the right moment – to listen or look too early or late in the year for the full chorus of songsters or the chicory bud unfolding.

We're all increasingly conscious of the pressures that climate change, urbanisation and industrialisation are placing on seasonal rhythms. In the UK, warmer springs mean plants are now flowering about a month earlier, on average. The risk is that this could push them out of synch with their pollinators, with worrying potential consequences for ecosystems and crop yields.[5] Yet, at the time of writing, actions to reduce emissions are falling short of what's needed to hit government targets. What's more, we're losing wildlife at a precipitous rate.

These kinds of headline stories have felt much more immediate and tangible to me in the process of sleuthing for signs of the times of day in my neighbourhood. I'm delighted at how well the melodious blackcaps are doing in my local park. But I remember the first year when the blackbirds failed to sing at dusk outside our building. I've come a little closer, too, to the sensitivity farmers and astronomers have to the impact of long spells of unseasonally wet weather: when plants fail, and the sky is blank.

I understand much better now the need to hold on to the knowledge of past generations and value the living world they knew and loved.

While I was developing this book, my sister Lizzie and I lost our Mum and Dad. And unexpectedly, the project became my own way of remembering them – keeping them close – by

INTRODUCTION

trying to share something of their appreciation of details in their surroundings.

When I was little, we'd often be out shepherding with Dad in the evening, looking carefully at what was happening in the fields. Or we'd be wandering in the garden with Mum, listening to the blackbird, admiring the flowers glowing in the soft early dusk. Mum might bend down a rose so we could feel the petals, admire its intricacy, draw in its scent. Often we'd lean beside her on the gate to the field and watch the towers of pink and red cloud stacking up in the west. Mum knew where the birds were nesting, watched for the swallows returning, and did an ace impression of a curlew. She was a wonderful storyteller, a great maker, and utterly alive to beauty. The summer after she died, the garden bloomed with luscious roses in unbearably, heartbreakingly lovely variety.

One Sunday in September, a few months after we lost Mum and when we knew we were about to lose Dad, I wandered through the farm as the Sun was rising. The village was sleeping. The thrush that had sung so vividly the previous dusk by the straw bales in the stackyard had gone. I walked to the old paddock and stood in the dew where I remember sitting as a small child in the grass drawing beetles and butterflies. When we were young – running or biking about in the fields – the ridges and furrows in this paddock were like a choppy sea. And now, in that quiet autumnal dawn, the waves seemed to have stilled.

I

Dawn Song and Bat Flit
Heeding the habits of fellow creatures

In the darkness, I can't see but I can *hear* the shape of the unfamiliar landscape. The jubilant crowing of the cockerel locates the farmyard behind me. The raspy cawing of the rookery is the wood beyond the house. The fluting of the robins and blackbirds are the garden and hedgerows to my left and right. The reedy piping of the coot is the river. And in the dim glow of first light, the trilling of the skylark rising and hovering in the quivering air is the silvery, dew-thick cow meadow.

The dark blue sky soon becomes luminous enough to see the ducks flying silently overhead. The pine, ash and oak, which moments ago were blunt, flat shapes, are tangling up with ivy. The pale clay track is unveiling its treacherous puddles. Milky drifts of mist are appearing in the valley like the aftermath of magic.

And while the brightening world starts to resume familiar form, the whistling hedgerows thicken with the songs of blue tits, goldfinches, chaffinches and dunnocks. An Old English word for daybreak is *dæg-wōma*, literally 'day-noise'.[1] In early March at Court Lodge Farm in Sussex (where we're staying

THE FULLNESS OF TIME

with Rosie's relatives), the *dæg-wōma* has gained singer after singer until – about twenty minutes before sunrise – it's in gloriously full voice.[2]

Is it possible to tell time by the sequence of birds joining the dawn chorus?

In the mid-1800s, several English-language journals broadcast the curious news from 'foreign journals' that 'a German woodsman has invented an ornithological clock'.[3] Before dawn in summer, according to the reports, the chaffinch – 'the earliest riser among all the feathery tribes' – starts singing from one-thirty to two o'clock and is followed at half-hourly intervals by the blackcap, the quail, the hedge sparrow, the blackbird, the lark, the 'black-headed titmouse' (coal tit?) and finally, from five to five-thirty, the sparrow.

'Queer horology that!' quipped one of the journals' editors. 'Quarter-past quail, half-past sparrow and about eleven minutes of blackbird.'[4] How absurd, indeed, to imagine wild birds giving time like the mechanical songsters on an ornamental clock. Naturally, living birds are not so strictly regimented. Dawn song depends on the season, the location, the variety of species present, and the impact of environmental change or habitat loss. The timing of the chorus moves earlier as the Sun rises earlier – and may be influenced by temperature, weather and moonlight.[5] And of course, birds don't pace themselves in precise half-hourly intervals. The dawn chorus I heard on the farm doesn't follow an identical pattern from day to day and month to month, even if I had managed to catch exactly when each songster started up.

Yet the general idea is borne out in experience: different species do join the dawn chorus in a broadly predictable sequence. Birds with better vision in low light are among the first to sing, like the large-eyed blackbird. There's 'a correlation with height', too, which may help explain why the skylark, rising up over the meadow, is another early singer.[6]

Rosie and I are visiting Court Lodge Farm on the warm invitation of her cousins, farmers Marian and David and their daughter Clare. Later today, over a fine Sunday lunch, they will introduce us to their friend Danny, a retired chestnut coppicer. When I ask Danny how the chorus sounds where he lives, he answers without a pause. 'It usually starts off with a blackbird – but sometimes a robin – just before it gets light. The blackbird really chortles away. Then, of course, all the others start coming in as it's getting lighter.' Most thrillingly of all, his beloved skylark sings.

Everyone here is deeply aware of the rhythms of all the creatures in their surroundings. (David has shown us with excitement the different moth species currently about on the farm.) For outdoor people, it's not a revelation that the dawn chorus generally builds up in layers as different species join in. I've always had a broad sense of it, because the birds I heard this morning are a familiar part of life on my family's farm. But only in the last few years (since Rosie and I made that van trip to the North) have I begun to understand the pleasures of waking early to fully absorb the dawn chorus. And there's nothing like the disorienting effect of standing in darkness in an unfamiliar place to quicken the senses and sharpen the ear.

I wonder why the German woodsman's 'ornithological clock' was considered newsworthy in the 1850s and 1860s. Perhaps the notion appealed to nineteenth-century tastes for cuckoo clocks and ideas about mechanistic, orderly nature. Or, possibly, it seemed delightful and curious because it appears to bridge two remote entities, the abstract calculative time of the clock and the vivid sensory qualities of time given by the rhythms of living creatures. For me – and maybe for readers in industrial cities in the nineteenth century – learning about the time kept by birds promises to draw us into a world of glorious song and vibrant patterns of life that feel remote in urbanised places.

In the brightening dawn on Court Lodge Farm, I walk back from the songful pasture to the handsome old barns with black corrugated roofs. In the corner of the yard, the rooster is crowing away in boisterous voice. He started long before I went out at the end of the night, and doesn't seem to be tiring.

The cockerel's bright call in the darkness lends its name to times of day around the world. In Old English, to give a local example, one part of the night after midnight and before dawn was named *han-crēd* (cockcrow). And there is a very old and widespread tradition that cockcrow marks successive times through the night.[7] In Ireland, for instance, a diary written in the early nineteenth century lists how country people divided the day and night: midnight is followed by 'first cock-crow', 'second cock-crow', then night draws to a close.[8]

Cockcrow could be a sound with profound significance. A rather lofty account given by Henry Bourne in the *Antiquities of the Common People*, published in England in 1725, says labourers distrusted going to work before cockcrow dispersed 'the Midnight Spirits'. 'Hence it is, that in Country-Places, where the Way of Life requires more early Labour, they always go chearfully to Work at that Time [cockcrow]; whereas if they are called abroad sooner, they are apt to imagine every Thing they see or hear, to be a wandring Ghost.'[9] The sound of cockcrow, it was believed, prompted malevolent spirits to fly away. But cockcrow could be untimely: to hear it at the wrong time of day was considered to be a deathly omen.[10]

The bird I heard rising in the dim light this morning – the lark – is 'messager of day' in Chaucer's *Knight's Tale* and 'herald of the morn' in Shakespeare's *Romeo & Juliet*.

In 'The Morning Quatrains' by the seventeenth-century poet Charles Cotton, after the cockerel crows to announce that 'the day's bright Harbinger', the dawn light, is peeping over 'the Eastern Hills':

> The merry Lark now takes her wings,
> And long'd-for days loud wellcome sings,
> Mounting her body out of sight,
> As if she meant to meet the light.[11]

Once the lark rises to welcome the day in Cotton's poem, the human household springs into action. The 'old Wife sits down to spin'. The milker takes her pail and goes to 'strip' Mull the cow's 'swoln and stradl'ing Paps'. The labourers and manufacturers set to work. The cultivator 'yokes his Oxen to the Team, / The Angler goes unto the stream', the woodman to the wood. The shepherd 'drives her Flocks, / All night safe folded from the Fox, / To flow'ry Downs'.

Cotton paints a dainty, idealising picture of English country people neatly divided by the labour they perform (and superimposed with classical imagery and symbolism). His poem suffers from the painful prejudices of his age. Yet Cotton plausibly describes a life where human routines are deeply entwined with those of other creatures and the cycle of the Sun. Where I stand now on this twenty-first-century farm, so much is different from Cotton's day. Yet tangible links remain between the rhythms of the birds and the rise and fall of daylight, and between the routines of people and cows.

When I first went out this morning, the beam of a torch was moving around at the other end of the farmyard. This was Archie, the young herdsman, getting ready for milking. The cockerel had been crowing for a good while before that, but today Archie was literally up with the lark.

Now the cows are waiting expectantly outside the milking parlour, where Archie is clattering about to set up the milking machine. He opens the parlour gate and adds his own calls and whistles to the songs of the birds and the lowing of the cows. 'Come on, girls! Whee-ip!'

Over the next hour and a half Archie will usher groups of cows in relay into the parlour. The milk needs to be ready for when the dairywomen arrive to make yogurt and labneh. Then, this afternoon, either Archie or Simon, the farm's senior herdsman, will bring the cows back for the second milking.

I didn't grow up in a dairy-farming community. And when I bump into Simon by the barn, I stop him for a chat between jobs. The aim, he tells me, is 'to try and get a nice level gap between the two milkings. So, you milk at five in the afternoon, and five in the morning.'

Simon has been milking for forty-eight years. 'It's a gorgeous life,' he says. And, I know, a very arduous one. But he's 'all in' and cares deeply for the animals in his charge, like David, Marian, Archie and almost every farmer I've known. 'I wouldn't stop it for the world,' Simon tells me. 'I mean, the winter could be a bit rough now. I'm getting on a bit. But the summer. Coming out in the summer, quarter to five, to get the cows in – it's just lovely. It's peaceful, you know. The wildlife here is just tenfold.' David and Marian converted the farm to full organic status in 1998 and rent land on the Sussex Wildlife Trust's Pevensey Marshes Nature Reserve. Along with all the birds I heard this morning, there are marsh harriers and bearded tits.

It's early March now. As summer draws nearer, sunrise will move earlier and the dawn song will follow. Whereas milking will continue on the farm here at more or less the same clock times.

Milking in the morning and evening is a very old pattern (though I'd speculate that, before the clock, the timing fluctuated as the days grew longer and shorter).[12] And there are likely to have been different milking times across places and eras, as there are today. A seventh-century Irish law text on keeping bees, *Bechbretha*, explains that the range covered by foraging bees is 'as far as a cow reaches in grazing before milking time'.[13] The word used for milking time, 'etrud', appears to mean midday.[14] In short, the bees travel the distance a cow walks in the morning. That measure of bee-flight by cow-walk gives a glimpse of just how familiar people were with their daily routines.

For people passing by a traditional dairy farm like this today, the sight and sound of a farmer calling and the cows gathering for milking still gives a broad sign of the time of day. And in the past, if not as much now, other sights and sounds in the working life of a farm would provide a familiar rhythm.

When shadows grow long, writes Charles Cotton in 'Evening Quatrains', the sheep are brought to the fold. The bees return to the hive. And the cock is cooped for the night, '[f]or he must call up all the rest'.[15]

THE FULLNESS OF TIME

That concluding event in the day of the rooster – his return to the roost – brings to mind an old name for a time of day: *cockshut*.[16] In *A Worlde of Wordes* (1598), John Florio translates the Italian expression *cane e lupo* as the time of 'cock-shut, or twilight, as when a man cannot discerne a dog from a Wolfe'.[17] The same phrase appears in French: *entre chien et loup*, between dog and wolf. Maybe it meant being caught between two kinds of threat. But I'm more convinced that the saying evokes the trickster state of twilight when it's hard to tell between guardian and predator, friend and foe, benign day and dangerous darkness.[18] No wonder wise chickens withdraw to the coop without prompting at cockshut.

In lowland Leicestershire, where I grew up, the sheep graze in hedged paddocks and are not penned in a fold overnight. But at shepherding time – usually after tea (the evening meal) and always before dark – Dad would go around the fields checking the flock was fit and well. He would be especially vigilant in the lambing season. Rotund ewes are sometimes prone to getting pinned upside down in hollows by their own weight. And a thick hedgerow is no deterrent to foxes sneaking in to snatch small lambs. (When the ewes and lambs were most vulnerable, they'd be brought into the barn.)

In wilder places, where you might actually be caught 'between dog and wolf', no doubt itinerant herders would be especially alert to the creeping approach of dusk. Nina Gockerell's rich essay 'Telling Time Without a Clock' (1980) reports that on overcast days the Hutsul shepherds of the Carpathian Mountains would observe their sheep's eye-pupils for a rough guide to the time. The shepherds 'knew that the pupils of sheep's eyes were oval throughout the day but became circular at just about the hour they were driven home to their folds'. According to Gockerell's source, Polish farmers similarly checked their cats' eyes.[19] Was the practice very widespread? 'The goats' eyes were my clock,' wrote the eighteenth-century diarist Ulrich Bräker of his youth as a herder in the high Swiss valleys.[20]

You might say I was a child shepherd, too, if you'd count the decidedly patchy efforts of a shell-suited teenager chasing down hedged lanes after her lost sheep. My sister Lizzie, who worked alongside Dad, has shepherded for most of her life. In contrast, I lack the temperament and the skill.

But after reading Gockerell's essay, I decided to see what happens to sheep's eyes for myself. One late afternoon in summer I wandered through the farm to see if anyone was about who could help me bring in the sheep and take a few photos. As I recall, a yellowhammer, perched on a telephone wire, was singing his distinctive appeal for a-little-bit-of-bread-and-no-cheeese. Somewhere in the distance, a tractor was moving around. But the yard and sheds were empty of people and sheepdogs. So I walked out alone into the field. With unexpected success, I shushed a dozen obliging ewes into a pen. Then I looked hard into their eyes.

I couldn't tell where the Sun was hiding behind the blanket of cloud and, although it was late in the day, it still felt fairly bright outside. At this stage the ewes' eye-pupils were a narrow pill shape, as you can see from my first photo, below.

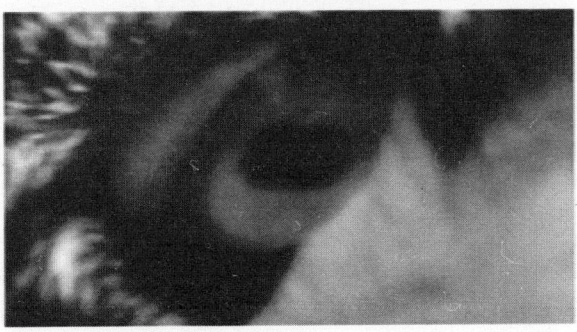

THE FULLNESS OF TIME

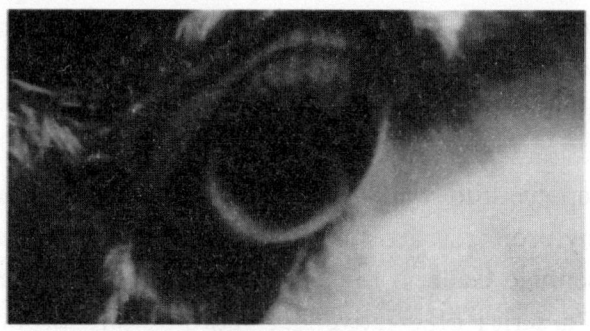

Around three-quarters of an hour later I came back from a walk to take more photos. Already, as the second picture shows, the ewes' pupils had morphed into round shapes.

It was quite startling to see these changes because I wasn't especially aware of how much the light had dimmed. Presumably that's partly because my own vision, too, was adapting to the failing light. By my subjective and unscientific field experiment, I reckon the sheep's eye-pupil test would be a useful prompt if you need all the signs you can get for the time to retreat for the night.

In Thomas Hardy's novel, *A Pair of Blue Eyes*, when Mrs Swancourt remarked on 'how truly people who have no clocks will tell the time of day', her husband agreed. He had known farm labourers, he said, 'who had framed complete systems of observation for that purpose':

> By means of shadows, winds, clouds, the movements of sheep and oxen, the singing of birds, the crowing of cocks, and a hundred other sights and sounds which people with watches in their pockets never know the existence of, they are able to pronounce within ten minutes of the hour almost at any required instant.

This sounds like a romantic exaggeration. Yet clearly the people who told time by the eyes of other creatures had a deep sensitivity for fleeting changes in the world around them.

That sheep-eye study was background research for *Stereochron Island*, my project as artist-in-residence in Victoria Park (for the Chisenhale Gallery) in east London in 2014. At its core, this was a series of creative field studies. My proposition was that we would imagine the park was a tiny island state, Stereochron, attempting to become a territory without clocks.

To support our mission, we invited experts to lead us islanders (interested members of the general public) to investigate signs of the time in Victoria Park from shadows, flowers, birds and other creatures. I'd heard somewhere that sheep had grazed here in the twentieth century – and in our playful thought experiment, we imagined bringing them back.

Our field studies began with a *gökotta*, as the Swedish call an early-morning outing to revel in birdsong. At 4.30 a.m. one morning in April 2014, a group of us from various parts of London gathered outside the park's locked gates with birder Peter Beckenham as our guide. The ranger opened up, and we plunged from the murky orange light of the street into the dense darkness.

With our vision suddenly reduced, the clarity of the songs of blackbirds, robins and wrens seemed enveloping, overwhelming. As we walked deeper into the park, the chorus swelled with the squeaky wheeze of great tits, the low coos of wood pigeons, the cascading trills of chaffinches, the dulcet whistles of blackcaps, the harsh caws of crows.

On the eastern skyline the pool of cold-grey light grew and yellowed until it threw an amber cast onto the treetops. A pink rainbow arched over the park. Big pink birds began circling – and after struggling with the sight for a moment, I realised they were gulls tinted by the rising Sun.

Our *gökotta* in Victoria Park was the first time I'd fully devoted my attention to the dawn chorus in London. There were no larks, but there were chaffinches, blackcaps and goldcrests! The range of inner-city birds was a shock.

In spring the early-morning air resounds with the clamour of 'love and war', as birders say. The male birds are flashing their ankles and raising their fists – appealing for mates and defending territory – through the medium of song. They will carry on singing through the day, though less consistently. In the heat of early afternoon there may be a lull. Then around sunset there is a quieter surge of sound: a dusk chorus or evensong.

Many daytime-active birds sing fastest and most fervently at dawn. But why they do so then remains an open question for scientists, with several possible explanations. In the darkness and half-light, the singers are masked from their predators. It's not yet bright enough to forage for seeds and insects. This is the time of day when female birds are most fertile. And when they typically lay their eggs, too, with the ability to receive a mate straight away.[21] Other theories are that the acoustic properties of the twilight atmosphere allow calls to travel further. Or that morning song is a kind of roll call of who's still here and holding territory after the treacherous night.

By mid-April, the dawn chorus is in full swing as millions of birds return to the British Isles from over-wintering in the warm south. Slight changes in the duration of daylight trigger a restless urge to migrate, known by the German word, *Zugunruhe*. At this time of year, each new wave of songsters arriving at their breeding ground adds to its chorus. After spring, some crepuscular (twilight-active) species like robins and thrushes will keep singing at dawn for much of the year. But once birds are nesting, as a whole their song subdues, becoming hushed after June, and picking up a little in autumn.[22]

In early autumn, migrating birds begin to fly south. Most birds make the journey at night (for reasons that are not all fully understood). In darkness, birds can evade their daytime predators. The nocturnal air may be calmer. It's cooler, too, which would protect them from overheating. And as they make their way through the night over land and sea, many birds are steering by the stars.[23]

What magical events are happening in the world beyond our perception! Gradually, though, after the *Stereochron* project, my appreciation for birds slipped away. Rosie and I moved to a noisier part of inner London, and life got more hectic.

But then came the 'great pause' of early 2020, when the pandemic lockdown stilled the machines and brought that unexpectedly joyful side-effect: the sudden intensity of the birdsong. Its beauty and clarity galvanised me not to neglect the birds around me quite so much again. And one of the great delights of this new commitment has been to become more aware of their vivid presence in the darker months.

In winter in the city where I live, it turns out the bare trees in the park are not empty: there are multi-species groups of small birds quietly moving about together. In the street I'm becoming more sensitive to the difference between the flicker of a falling leaf and the darting of a tiny wren. And of course not all species are silent in the dark months. At dawn and dusk I've heard the blackbird singing its gloriously complicated song in the bare branches of a cherry tree.

Out in the winter countryside, as the Sun sinks, scattered groups of starlings gather into a great body of bodies that swirls in synchronous motion and murmurs with the fluttering of thousands of wings. Herons and egrets roost in the trees like ghostly lanterns. The solitary white phantom – the barn, screech or ghost owl (*Tyto alba*) – swoops silently over the darkening fields with loping wings, or calls unseen with hair-raising shrieks.

An echo remains in our everyday speech of how closely people once tuned into birds: we still say we're 'up with the lark' to mean early morning. Whether or not past generations had practical need to tell time by the cock crowing or the songsters quietening, the rhythms of birds would have given shape to the day and seasons. For us now, I think, there is personal bounty to be gained in becoming a trill seeker – and especially to listen closely to the avian world as the darkness ebbs and the light fades. The more I notice in the dead of winter as well as spring, the more I feel the turn of the day and year from light to darkness and back again.

After the Sun has 'gone to sleep' in Cotton's gloomy poem 'Night Quatrains', the owl flies, the nightingale sings, the bittern 'booms'.[24] Over the ages, these and some of the other twilight- and night-active creatures he describes have suffered declines or vanished almost entirely from our surroundings and awareness. The nightingale, not least, is only to be heard in rare places in the countryside. But I'm delighted to learn that a small number of bitterns have found a winter haven at the London Wetland Centre. There are owls in the great city parks too. And as we're about to discover, even in some of the most unlikely places, the night air is not empty of life.

It's a warm evening in early June and I'm in Feltham Marshalling Yard, a stretch of abandoned railway land in west London just a mile or two from Heathrow Airport.[25]

A deliriously lovely evensong is rising up from the blackcap in the bramble and the song thrush in the willow – despite the overwhelming, body-rumbling thundering of air traffic. The direction of the wind this evening means aeroplanes are roaring up into the sky right overhead.

A group of us have gathered for a public walk to see bats. As a whole we're curious but not knowledgeable. And I suspect we're all intrigued by the prospect of experiencing this iconic nocturnal animal in an urban wasteland so close to an international airport.

Our guide is the conservationist Elliot Newton from Citizen Zoo. He hands out heterodyne bat detectors: small grey plastic boxes with knobs and dials that have the air of an eighties TV sci-fi prop. We twiddle with the settings and produce a weird chorus of mechanical buzzes and whistles as if we're sending signals into the cosmos. Someone jokes, 'The detectors aren't going to interfere with the aeroplanes, are they?'

The low Sun throws pink-golden light over the scrubland around us: twilight is approaching. Elliot encourages us not to use torches unless we really need them, but to let our eyes adjust to the growing darkness. As for the bats, they'll come out when the light, temperature and weather conditions are right. And they won't mind our presence, he assures us. On this warm dry evening we can expect some bats to start flying around fifteen minutes after sunset. Then they'll hunt until dawn, returning now and then to the roost to rest.

We pick our way along a path through the brambles into the woodland. Elliot brings us to stand in an oak glade beside a burnt-out moped. The trees are strung with orange and pink lights, which I slowly realise are fragments of sunset sky shining through the foliage. A cloud of midges dances in the warm pocket of air trapped under the leaf canopy. Midges – along with moths, mosquitos, lacewings and other invertebrates – provide bats with their main meal. Despite alarming declines in invertebrates, Elliot says, currently eleven of the UK's seventeen breeding species of bat seem to be doing okay.[26] He reckons this glade is a likely spot for hungry bats. Those we can hope to see are the pipistrelles, whose acrobatic flight concentrates on the mid-canopy insects. And the much larger noctule, which flies above the trees to hunt the high-flying invertebrates, like winged ants, moths and flying beetles. If we're lucky, we might come across a Daubenton's bat by the river, skimming to scoop waterborne insects, like caddisflies and mayflies, with its net-like feet.

'Here we go!' Elliot pivots and swivels with his detector, picking up the fast clicks and chocks of a soprano pipistrelle that we can just glimpse racing through the glade.

Contrary to the folklore, the bat is not blind: the retina in its eye is densely packed with rod cells that reveal objects in darkness.[27] But the bat's main sensory perception of its environment is through the sound waves reaching its large, sensitive ears. As the bat flits through the air, it maps three-dimensional space by the sound of its own voice, with a technique known as echolocation. As the bat beats its wings down, Elliot tells us, the pressure adds force to the cry from its throat. The bat interprets the timing and quality of the returning echo to judge its own position relative to objects around it, as well as the location of its prey.

The flitter-mouse or flicker-mouse, to call the bat by vivid old names, flickers at the edge of our vision in the darkness. If there were a child among us, they might be able to hear the bottom range of its cry unaided. But most adults can't hear it without the bat detectors, which bring down the pitch of the bat's voice by a few notches.

Imagine a scream coming at you louder than the planes thundering overhead. I wonder idly if that sensation might give us some idea of how the bat's voice trembles through a moth's body. Though whatever this sweeping, twisting pipistrelle is chasing, it seems to be putting up a good fight.

Noise is powerful and disruptive, as we've all experienced. In response to intense urban din, some birds in some places may be singing earlier and at higher frequencies.[28] And when bats hunt by rumbling highways, unsurprisingly, there is evidence to suggest some species may hunt less successfully.[29] I am far more oblivious, though, to the silent risk posed by another urban peril we're about to encounter.

In the dim blue light of dusk, Elliot leads us in single file around the edge of the dark wood. The murky shadows have drowned everything but the white ox-eye daisies, creamy elderflowers and silver birches, which seem faintly aglow. The spherical seedheads of the goat's beard plants at our feet are like cobweb lamps along a fairy track.

Suddenly the path bursts from the wood into the blazing light of an empty train depot. The glare leaks out and up beyond its boundary. The colours and forms of flowers and trees have been restored to slightly surreal versions of their daytime selves. Inside the depot, all is still and quiet. Except every lamp is humming with insects, like a buffet bar piled high for predators.

What need is there for such an intense light spill, Elliot asks. These insect-luring lamps may look like a boon for a bat on the hunt. But, he says, the bright glow presents one of the biggest dangers to them, too, especially species such as

Daubenton's bats. To reach their foraging grounds without flying into lit areas, they may be forced to take more difficult routes that are longer, more tiring or more hazardous.[30]

Artificial light, of course, extends the day – and that poses another potential problem for some species. Bats not only roost in caves and old woodpecker holes but within built structures like barns, bridges and churches (which might actually have 'bats in the belfry').[31] I live near a fine old building and was about to propose illuminating the tower at night. But if floodlamps are trained onto a roost, that could delay when its inhabitants emerge in the evening. They may miss the time when insects are most abundant around dusk. And when artificial light spills inside their home, there's even a chance they may fail to hunt and become 'entombed' in the roost.

Elliot leads us back through the shadowy wood to a bridge over a river, the last stop on our walk. This place is a major bat corridor, he says – and as if on cue, our detectors start spiking with jagged pops and squeaks. The sound adds to the adrenaline rush. But I switch off my detector and stand in stillness watching the flicker-mice swerve and dart against the cobalt sky. Their small bodies are slipping softly through a parallel world of sounds masked from our human senses.

On the walk back through the streets to the station, the scent of roses hangs in the warm air. Heavy traffic growls on the Uxbridge Road. And I wonder if any travellers jetting up into the sky overhead are imagining the aerial marvels being performed in the woodland below. An international airport is a kind of time-defying 'portal' from which we leap from the present on the ground and dart between time zones. How different the airport's rhythms seem from the daily patterns of birdsong and bat flight at its edges.

I head back into London. The windows of the brightly lit train carriage are black mirrors animated with floating patterns of yellow and white lamplight. The world outside seems blank, empty, irrelevant. But I'm newly aware of how the air is full of life. There are bats wheeling above the streets and moths trying to dodge them. And in two or three months, migrating birds will be steering their way south through the night over land and sea.

The train runs into the centre of the city, where the brighter light erodes the darkness. I guess my eyes' pupils are narrow now, and I wonder how other creatures are responding to long-term exposure to this nocturnal environment. Later I will read about initial evidence from the US that suggests some city-resident birds might be developing smaller eyes in response to light pollution.[32]

Notoriously, too, there are cases where migrating birds become fatally disoriented by the lights in tall buildings and crash into them. Fortunately, in some cities around the world, corporations are switching off their lights, particularly during the bird migration seasons.[33]

At home I turn off the lamp in my backyard and allow my senses to adapt to the urban semi-darkness. There are bats here, too, swerving over the honeysuckle, chasing moths.

Traditionally the routines of fellow creatures gave shape and quality to the human day. Even in the twenty-first-century urbanised West we still associate the crowing of the cockerel with dawn, the humming of bees with summer days, the swooping of the bat or the hooting of the owl with the night. But where I live, cows, cockerels and owls seem most present in stories or movies and on cereal packets or billboards. They are narrative devices, mythological symbols, brand logos, cartoon characters.

THE FULLNESS OF TIME

For a long time I thought I might be the first person on the farming side of my immediate family not to live with other animals – and consequently not to gain a sense of the rhythm of the day from their presence. My inner-city neighbourhood feels a world away from the lively fields, woods and hedgerows on my family's farm and at Court Lodge Farm. But even here, when I listen and watch carefully, I'm growing increasingly aware of the more-than-human life around me. My sense of the time of day is enriched by the dawn songsters and the buzzing bees. I'm more alive to the moment in the evening when the blurry presence flitting past the window is not a sparrow but a bat.

By comparison with these darting beings, the flowers in pots by the door are models of static grace. Yet, as we're about to discover, some flowers have predictable daily movements that are wonderfully tangible to us.

2

Day's-Eyes and Turnsoles
Following the hours of flowers

On warm summer evenings in the streets where I live, the sweet-sharp scent of honeysuckle drifts from vines tumbling over garden fences. The seventeenth-century London diarist Samuel Pepys is said to have called these tubular blooms 'ivory bugles' that 'blow scent instead of sound'. Today their dusk perfume is still a palpable signal that the day is ending.

When we think about how plants mark time for us, it's obvious to picture the pattern of the year through botanical events: flowers springing up in spring, fruit ripening in summer, leaves falling in the autumn, and it all dying back in winter. But how easy is it to picture flowers heralding times of day? If you'd asked me this some years ago, I would probably have thought of the sunflower, whose head is renowned for following the Sun. And of course there are the plants like common honeysuckle* and jasmine,* whose scent usually grows more intense in the evening. But aside from that I might have drawn a blank – until a curious plant encouraged

* *Lonicera periclymenum.*
* *Jasminum officinale.*

me to become more entranced by flowers with expressive daily movements.

This was in the early 2000s, and Rosie and I had just moved into our first home with a small garden. One of the flowers I planted there was a creamy yellow evening primrose,* which (if I remember rightly) put out a delicate nocturnal fragrance. But the great allure for me was that, at the end of each day, the plant produced a fresh bloom. Evening after evening we'd sit and watch the crumpled yellow petals of an infant flower break free from the jaws of their pink casing.

Over the years we lived alongside this plant and its offspring. And eventually, I made a video of a new evening primrose unfolding in our backyard. The whole event took maybe twenty or thirty minutes. But the signal movements happened very quickly: the pictures, above, are stills from the video taken four minutes apart.

* *Oenothera.*

As the flower matured, the pale lemon petals would slowly turn orange-pink, giving another subtle sign of the passage of time. They would slump and dry like a limp rag and fall away, then gradually be replaced by an addition to the seedpods slowly accumulating up the stem.

I was completely seduced by this plant. It was the first flower with regular rhythmic movements that I got to know. In the clock-bound city – where the daily grind can feel so unrelenting and unchanging – the release of fragile new petals into the world as the light dimmed gave a very different and somehow more ample quality of time.

When Rosie and I moved to another part of London, we left our evening primroses behind. But by then I had been researching other flowers with animated daily routines. And I'd been surprised to discover we come across one of them all the time here: the common daisy,* whose name comes from the Old English for day's eye.[1] People call it that 'for good reason', as the fourteenth-century poet Geoffrey Chaucer remarked.[2] The little white day's eye does indeed open in the morning and shut at the end of the day (and in poor weather). This 'sleep movement' in plants is known by the pleasing scientific name of nyctinasty.

The daisy is in the family *Asteraceae* and shares the habit of unfolding and folding at predictable times of day with some other members of the family, like the common marigold* and the dandelion*.[3] I took these pictures, below, of a pair of dandelions on a summer's day at midday and in the afternoon and evening.[4]

* *Bellis perennis.*
* *Calendula officinalis.*
* *Taraxacum* species.

Dandelions, like daisies, are scattered everywhere there's turf or gaps in the paving. And over the years I've become especially drawn to their bright yellow tasselled discs. It's particularly heartening to notice on a lunchtime stroll how the dandelion flower-heads are pressed wide open like tiny suns.

Even in our new neighbourhood, with its dense maze of tarmacked streets, there are hundreds and thousands of dandelions and day's-eyes blinking in slow motion. And in recent years the daily rhythm of another old bloom I'd associated only with the countryside – the common chicory* or succory – has become more and more present.

Our local park is bordered on one side by a line of small artificial hills that vaguely resemble prehistoric long barrows. Over and around these mounds, the gardeners have sown a flower meadow. Despite the busy roads right beside them, here crickets sing, bees hum and butterflies flutter by. In late summer the dry meadow plants form a thick carpet the colour of burnt butter, from which spring sky-blue chicory flowers, feathery fronds of mugwort, fruiting heads of cow parsley. If you wander off the path, you'll be picking seeds from your socks.

After a few summers the chicory blooms grew so profuse and tall on their stalks that they hovered above these miniature hillsides and plains like a pale blue mist. One evening

---

* *Cichorium intybus.*

I bumped into my friend and neighbour Jmeel, who'd been admiring this great drift of flowers which looked, he thought, 'like a reflection of the sky'. He'd noticed it earlier in the day and had just gone back to take photos in the golden light. And he twinkled at me now because he knew I'd like this: he'd arrived to find the spectacular haze of blue speckles had vanished like an optical trick. The chicory flowers had closed up.

Jmeel was aware of my fascination with vividly animated plants. He's married to my friend Shab, whom I'd recently given a purple shamrock* with large triangular leaves that fold themselves down for the night. And there was the episode three years earlier when Shab had bumped into me hovering with a camera by a shrub in the carpark outside our building. Weirdly, she wasn't surprised to find me lurking there. It turned out Jmeel had been doing the same thing. We had both spotted an intriguing plant that had miraculously seeded itself in a neglected island of soil by the bike shed, maybe having floated over from the park meadow. This was a goat's beard,* a plant with a large and splendid spherical seedhead that appears to be made of tiny whiskery shuttlecocks. (We saw its pale globe in the dusk when we were looking for bats in the last chapter.) In comparison, the flower-head was less remarkable. From a distance, an unkind critic might have called it a shy-looking little dandelion on a lanky stem. But it was the bloom, not the seedhead, I was most excited to see. Because, regardless of its subtle presence, behold what the flower-head did in the middle of the day.

* *Oxalis triangularis.* (The flowers close too.)
* *Tragopogon pratensis.*

THE FULLNESS OF TIME

 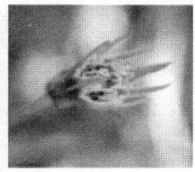  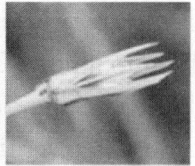

The first photo, above, pictures the goat's beard in late morning. The next three show what happened over about quarter of an hour shortly before the time the Sun was highest in the sky. At this stage the green bracts were closing around the yellow florets quite rapidly, before slowing down until the bloom seemed deeply asleep by mid-afternoon. (It goes without saying this was amateur botany and not a scientific study, not least because the plant was in a shady spot and bobbing in the breeze.)

Aptly, the goat's beard's other names include Jack-go-to-bed-at-noon, noontide and (my favourite) nap-at-noon.[5] And while the plant was in flower, whenever I rushed out for a meeting or zipped home for lunch, it was vivifying to see where it was at in the course of its own day.

Over time I've become more alive to the opening and closing times of other local botanical beings, like the yolky yellow courgette flowers yawning wide in the allotments early in the morning and a neighbour's pink-and-yellow four o'clocks* waking in late afternoon. It seems to me that once you start noticing the population of these special kinds of plant in your area – when you're on winking terms with your local daisies, say, and sensitive to the habits of the creeping wood sorrel* in a street gutter – the world starts appearing with a new filter. A flowerbed is not only a source of sensory pleasure but a site of intrigue. Even a straggly

* *Mirabilis jalapa.*
* *Oxalis corniculata.*

clump of opportunists in a pavement crack is no longer something to ignore. Once you're alive to the habits of marigolds and all the rest, they leap out from the crowd, and soon you're seeing 'time' wherever there are flowers and leaves.

Is this how people in a world without clocks felt about those special blooms? It is tempting to imagine that the day would be patterned by drifts of time-giving flowers spreading through yard and field and running beside you as you walk. And that the routines of plants would be part of the wider weave of daily time signals, along with shortening shadows and the evening star.

Traditional names for some plants describe their vivid daily routine, like Jack-go-to-bed-at-noon. The dandelion's local names include shepherd's clock, farmer's clock and time-teller – and I suspect they don't all come from the tradition of calling its seedhead a 'clock'.[6] The scarlet pimpernel,* whose little flowers open in the morning and close in the early afternoon (and in dull weather), was also known as the shepherd's clock.[7] Certain names appear to confirm their daily habits were noticed.[8] But did people actually take signs of the time from this kind of lively plant with predictable routines?

One answer comes from Pliny the Elder two millennia ago, when he pictures Nature proclaiming to the husbandman that she has given him plants that show the hours (*horarum indices*).[9] True, Pliny is not giving an eyewitness account of cultivators telling time by plants in the Roman fields. But I have caught two potential glimpses of country people in the British Isles referring to time-telling flowers. The Elizabethan gardening authority Thomas Hill wrote that the marigold is called the Husbandman's Dial because the 'opening and

---

* *Anagallis arvensis.*

shutting' of its flower 'declareth the houres of mornyng and euening'.¹⁰ And *The Rural Cyclopedia*, published in Edinburgh and London in the 1840s, mentions that 'the farm-boys of some districts regulate their dinner-time by the closing' of the goat's beard.¹¹

Such reports are tormentingly scant. And it may well be that it wasn't common practice, at least in the British Isles, to look down at dandelions or marigolds for a sense of how long to go before sunset. Or perhaps, like many features of everyday life which seem too obvious to record at the time, flowers with rhythmic movements were once a minor part of the mix of everyday time-tellers out in the countryside.

Whatever the case, it's not unreasonable to expect people's awareness of the hours of flowers would fall away in Europe in the seventeenth and eighteenth centuries when mechanical timepieces were becoming more accurate and ubiquitous. Maybe that is what happened in ordinary life. Yet this is the era that produced the two most famous timepieces made of living flowers in European history.

The first of these legendary clocks was inspired by the mighty sunflower (*Helianthus annuus*). For many of us around the world today, there can be few more recognisable botanical beings than this tall plant with its huge flower-head ringed with gold. But in its early years in Europe, the sunflower was a rare marvel with a head-turning reputation.

The sunflower had been cultivated in the Americas for thousands of years before its seeds were taken by colonisers and explorers to Europe in the mid-sixteenth century. At the end of that century, the English herbalist John Gerard shared with readers his experience of growing a 'Sunne flower'.¹² This plant is 'of such stature and talenesse', he wrote, that over just one summer 'it hath risen up to the height of fourteene foote in my garden'.

Gerard's garden is thought to have been on a site by Fetter Lane in Holborn in central London, which was then a suburb. When I last cycled down the lane, a couple of dandelions had not yet been zapped from the pavement, and a 'green wall' of plastic plants adorned a bar front, but there were few botanical surprises. Imagine, though, passing through about four centuries ago. If you'd never heard of the Sunne flower, how impressive would it be to encounter this resplendent plant towering above Gerard's wall? He described the flower-head as 'sixteene inches broade', with a 'border of goodly yellowe leaves' (ray florets). The 'middle part', he wrote, 'is made as it were of unshorne velvet'.

Gerard was inclined to think the flower of the Sunne ('*Flos Solis*') 'was so called bicause it doth resemble the radiant beames of the Sunne', and not, he implied, because of its reputation for following it. And when he lists the sunflower's several other monikers, he doesn't include 'heliotrope' or 'turnsole', which were given to plants like the marigold and sun-spurge* that were believed to or did turn through the day with the Sun. Yet others, perhaps later, would call the sunflower by those names,[13] and in French the plant is still called *tournesol*.

The marigold had long been a symbol of fidelity for its devotion to the Sun. And it's no surprise to learn that the huge and majestic sunflower in full bloom, which offers so much dramatic possibility, attracted a panoply of meanings. In seventeenth-century portrait paintings and emblem books (collections of allegorical images and mottos), the sunflower could represent profane love and loyalty. In religious imagery,

---

* *Euphorbia helioscopia*.

the sunflower – turning with the rays of the Sun – illustrated the Christian soul's orientation to the divine light. Or the sunlike sunflower stood for the day and the opium poppy the night, with the pair representing the opposition of good and evil. This mighty plant, it seems, could bear weighty meaning.[14]

One of the most spectacular ideas of the age was that the splendid sunflower could be made into a working timepiece. The inventor of that wondrous device was Athanasius Kircher (1602–80), a German Jesuit priest and polymath of extraordinary range.

Over his career he would be the first, for example, to propose disease could be caused by germs (rather than, say, devilry) and to produce an entire book on the analysis of acoustics.[15] And he would invent countless theatrical devices: intricately complex sundials and clockwork timepieces, an astrological moondial, an organ reputed to play all varieties of birdsong, a contraption with mirrors that magics up spectres in mid-air, hydraulic statues that vomit up drinks to delight guests. In later life Kircher and his museum of marvels would be sought out by philosophers, popes and princes.[16]

In 1641 he presented a fine illustration of the sunflower clock in his masterwork on magnetism.[17] The plant is pictured in full bloom, floating in water on a cork base, inside a circular band marked with hours. A pin in the flower-head points to the time through day and night while it fully rotates by (what Kircher theorised to be) the magnetic force of the Sun.

The sunflower clock, however, could not perform this time-telling trick without hidden magnets. Indeed, it would be just one of Kircher's grand claims and thrilling inventions that would not bear close scrutiny. And over his career, this great luminary would be mocked by peers for his love of trickery and spectacle. Yet he had a deeper purpose than simply to entertain. As the historian Mark Waddell told me, Kircher presented his sunflower clock as visible proof of what he believed to be the invisible natural forces that connect myriad things in the universe.[18]

Had Gerard lived long enough to see a picture of Kircher's clock, I suspect he would have been quizzical. In his notes on the sunflower's names, he remarked that some 'have reported it to turne with the sunne, the which I coulde never observe, although I have endevored to finde out the truth of it'. The mature sunflower has never quite shaken off that reputation for following the Sun, yet Gerard was right. Only the *immature* bud rotates, turning slowly west through the day, then retracing its path overnight. Once blossomed, the flower-head stays rigidly facing the place where the Sun rises for the rest of its life.[19]

One recent August I admired a dwarf sunflower that someone had managed to nurture in a bare channel of earth by a local supermarket. It looked so vivacious and abundant against the parched soil and brick, with its huge flower-head basking in the warmth of the morning Sun. The florets were so loaded with powdery yellow pollen that it fairly caked a visiting honeybee and had shaken down onto the leaves. When I passed by in late afternoon, the sunflower was still looking east. Only its shadow had been turning through the day.

The captivating idea of telling time by living flowers did not fade. Some years after Kircher, the English poet Andrew Marvell described a garden where 'sweet and wholesome hours' are 'reckoned' by a 'dial' formed from 'flow'rs and herbs'.[20] And in the next century a new model scheme was proposed for a working clock made of many flowers, which has captured imaginations ever since.

Picture a garden gleaming softly at the end of a short summer night. A ring of numbers marked on the ground is surrounded by a bed of sleeping flowers. After a few hours of stillness, one by one, the flowers slowly stir. Between four and five, the sky-blue chicory buds peek at the day like waking eyes. Soon after six, the tight knots of the hawkweed* slowly burst into little yellow suns. After seven, the white petals of the waterlily* unclasp to show the golden crown within. At eight, the scarlet pimpernel reveals its tiny face. Between nine and ten, the starry flowers of the glistening 'ice-plant'* unfold as if to bask in the light. And on goes the slow-motion ripple around the ring as the opening of a different bloom marks each hour like a silently chiming bell.[21]

The notion that a kind of working clock could be formed from many blooms with regular rhythmic movements is credited to the great Swedish biologist and physician Carl Linnaeus (1707–78). In his book *Philosophia Botanica* (1751) he proposes floral clocks for any climate 'are to be worked out ... so that anyone can make a calculation of the hour of the day without a clock or sunshine'.[22]

Linnaeus may be best remembered today as the first to define simple scientific principles for naming plants and animals, known as binomial nomenclature. We still use this method of identifying each species by a two-part Latin name that combines its genus with a specific epithet. For example,

---

* *Hieracium umbellatum.*
* *Nymphaea alba.*
* *Mesembryanthemum crystallinum.*

Homo sapiens – as Linnaeus was the first to call us – is the wise or knowledgeable species in the genus *Homo*.

In *Philosophia* Linnaeus takes a similar approach to the flowers that open and close through the day, which are, he says, 'very well known'. He marshals them into tidy order, names them 'solar flowers', and divides them into three kinds. The first keep somewhat irregular hours because their times of opening and shutting are influenced by shade or atmospheric changes. The second alter their hours of opening and closing as the day grows longer or shorter. The third, which he calls equinoctial, 'open at a certain definite hour of the day, and for the most part also close up at a fixed hour each day'. The floral clock, Linnaeus says, 'should be formed' from the equinoctial flowers, and he provides a table of more than forty examples and the hours when they open and shut in his part of the world.[23]

Linnaeus was said to have 'devised his so-called Floral Clock on the basis of his long-continued observations at' Uppsala.[24] In 1741 he'd been made professor of medicine at the university, a post that came with a house in the corner of the botanic garden. From there he kept a very keen eye on the plants in the beds and orangery, while those under closest scrutiny were perched in pots on his windowsill. In summer, he recalled, he slept briefly and rose at 3 a.m., then rested for long hours on winter nights.[25] He sounds like a man who lived in synch with his plants.

I'd imagined Linnaeus in the botanic garden regimenting his specimens into a clock dial like my fantasy of a circle of waking flowers all taken from his floral clock table. I'd envisaged him testing plants for accuracy, removing poor performers, scrawling notes and finally standing with satisfaction amid his botanical timepiece.

But there is no evidence that he actually grew the floral clock[26] that's so vividly suggested by his table of plants, which don't all flower together in the same season or keep such tidy timings. Indeed it's thought that Linnaeus intended the clock to be a guide to time out in the countryside – and that the idea to plant it in a flower bed came later from one of his students.[27] What's more, it seems Linnaeus lacked the time to refine his hypothesis, and deferred further research to his son, who himself never completed the study.[28] Whereas his followers – and many others since – attempted to grow and perfect the scheme, recording the hours of flowers in other places and adding blooms that open in darkness.

By the late eighteenth century, the British affluent classes were increasingly beguiled by all things botanical, from plant studies to flower gardens and, apparently, even horticultural hairstyles.[29] Plants with theatrical forms and rapidly animated habits – some extracted from overseas and newly arrived in Europe – were a source of particular fascination. They are vividly portrayed, not least, in *The Loves of the Plants* (1789)[30] by the great polymath Erasmus Darwin (grandfather of Charles). This elaborate poem is a botanical 'soap opera'[31] set in a magical garden over the course of a day. And it was inspired by Linnaeus's system of classifying plants.

Controversially, Linnaeus argued that plants procreate sexually. Indeed, to the gnashing of traditionalists, his method of ordering plants was based on counting the reproductive organs inside their flowers. He called these organs 'stamens' and 'pistils' but also personified them as 'husbands' and 'wives' who procreate on the 'nuptial bed' of the flower's calyx – often in non-monogamous groups.[32]

While protectors of 'female modesty' would attempt to close the literary blinds on such distasteful goings-on in the garden,[33] for Erasmus Darwin the sex of plants was fundamental to Linnaeus's scheme – and hugely creatively suggestive.[34] Not only did he lead the production of translations of Linnaeus's work into English but composed this long poem (with copious notes), *The Loves of the Plants*, to bring his botany vividly to life for a popular – significantly female – audience.

In Erasmus Darwin's playful confection, personified pistils and stamens 'crowd the gaudy groves' and 'woo and win their vegetable Loves',[35] often in groups of multiple partners. For his cast of plant-women, the poet wheels out an eye-rollingly predictable pageant of modest maidens and dangerous vamps. When the night-blooming cactus *Cerea* blossoms 'at the dusky hour', she opens her 'fair lips, and breathes her virgin vows'. Later her 'polish'd bosom gleams' in the moonlight while she's gazed at by *'crowds'* of 'admiring swains'[36] ... and so on, in a style that may seem wonderfully camp or thoroughly disagreeable to us now. In its own age, the writer and connoisseur Horace Walpole declared it 'the most delicious poem on earth'.[37]

In the botanic garden in *Loves*, some players – like *Cerea*, who's modelled on a cactus with flowers that usually bloom for just one night* – have rapid and dramatic movements. And not all of these expressive characters are wooing or being wooed. *Lapsana* (nipplewort), *Nymphaea* (waterlily) and *Calendula* (marigold) – three female plants explicitly drawn from Linnaeus's floral clock – 'trace with mimic art the march of Time'.[38]

---

* The cactus now named *Selenicereus grandiflorus*.

The Loves of the Plants was a bestseller – presumably not just because of its spectacular and inflammatory verse, but for its treasury of notes on botany and many other themes. The mood turned in the 1790s, however, during the French Revolution, when thinkers like Erasmus Darwin were condemned for ideas that threatened the stability of the social, sexual and religious order. Artistic ideals changed too, and his literary as well as wider reputation collapsed.[39]

Nevertheless, lively flowers with rhythmic movements would keep springing up with brio in different poetic guises. One popular nineteenth-century poem, 'The Dial of Flowers' (1828) by Felicia Hemans, seems to look at Marvell's 'The Garden' and Linnaeus's floral clock as if to a lost and uncertain Eden: "Twas a lovely thought to mark the hours … By the opening and the folding flowers … To such sweet signs might the time have flow'd'.[40] I perceive echoes, too, of earlier poems in Tennyson's famous lines, 'Now sleeps the crimson petal, now the white … Now folds the lily all her sweetness up.'[41]

The 'famous dial of flowers … kept Linnaeus's name fragrant' for future generations, observed the late-Victorian writer Florence Caddy. 'The notion' of it, she wrote, 'tickled the English people vastly as it got into the popular books'.[42] And while scientific interest in it may have peaked by the late-nineteenth century,[43] the idea of sensing time by the motion of living herbs and flowers appeals to us still.

On a frosty November day, I went to meet the plant biologist Alex Webb at Cambridge University Botanic Garden. We strolled through the magnificent trees to find two crescent-shaped summer plots that Alex designed with the gardeners.

Species that mostly open in the morning are planted in one crescent, and the evening in its opposite. Together they form a day-and-night garden called the Circadian Beds.

When we arrived, we found the gardeners had just finished clearing the beds, and the ragged remains of last year's plants were heaped high in their wagon. But the rich, dark loam was full of promise. And with excitement I spied the pile of plant tags and information boards beside the beds. There were labels for an evening primrose and several varieties of *Silene* flowers, like the bladder campion,* which releases its clove-scented fragrance in the evening. On the boards were pictures of a stripy morning glory,* which opens at dawn and closes at the end of the day, and a purple relative* of Jack-go-to-bed-at-noon.

The Circadian Beds are a living exhibit enticing visitors like me to explore why and how plants keep daily rhythms. Obviously green life is not nimble in space. Even the rapid movers are rooted to the spot. But botanical beings are deft time managers, in that they have an internal 'clock' that coordinates multiple processes. Alex explained that a plant has two main ways of adapting to the rhythms of the world around it. The plant's sensory system responds to changes in the moment, like the dimming dusk or the gathering frost. And its inbuilt clock prepares the plant for what's going to happen.

Tomorrow morning, the Sun will rise at a different time from today. For a green organism, Alex said, this shift continually alters its opportunity to make energy through photosynthesis. (Photosynthesis is the process through which

---

* *Silene vulgaris.*
* *Ipomoea.*
* *Tragopogon porrifolius.*

plants manufacture their energy in the form of sugars from sunlight, water and carbon dioxide.) If a plant only *reacted* to the first glow of dawn, it would lose time. But its internal clock anticipates morning twilight and prompts its photosynthetic apparatus to set up, ready to absorb every sunbeam. That's one critical example of how the clock coordinates processes to happen at the right time.[44]

The dawn light resets the plants' clocks each morning, keeping them in synch with the cycle of sunlight and darkness in the world around them. Over the rest of the day and night, the clock also takes time cues from changes in the quality of the light, the ambient temperature and rising sugar levels as the plant makes energy from sunlight. This subtle timekeeping system matters because, as Alex's research group at the university has found, when a plant struggles to get in rhythm with the daylight and darkness, it will be smaller and weaker than its peers.[45]

To be clear, all plants have internal clocks, not just the crowd in the Circadian Beds. Indeed, almost all living organisms are now understood to have forms of biological clock. We call this timekeeping system *circadian*, from the Latin for 'about' and 'a day', because it's stretchy. This is evident when a living organism is confined in unbroken darkness for a few days and their biological clock cycle starts to run free, falling out of synch with the rhythm of the twenty-four-hour day outside. When the organism is brought back out into the world, the clock 'entrains' to its new environment, resynching with the local cycle of daylight and darkness.

The first scientific evidence that the internal 'clock' exists was indicated by experiments performed on a plant three centuries ago. This is believed to have been the startling and easily startled *Mimosa pudica*, known as the sensitive plant. If you were to tap one of its soft fronds, typically you'd find its rows of paired leaves fold in together remarkably quickly – almost, I think, as if slowly squirming. The leaves seem just as sensitive to nightfall, because at the end of the day they close together until the morning.

Back in the 1720s the French astronomer Jean-Jacques d'Ortous de Mairan confined the plant to constant darkness and observed its leaves continuing to close and open, despite not experiencing dusk or dawn. De Mairan doesn't seem to have thought the plant capable of moving without some kind of immediate cue from outside – but others did. Some used Linnaeus's term 'sleep' to describe plants closing up to argue that the sensitive plant does have a form of agency.[46] Much debate and research followed, with the most significant experiment coming about a century after de Mairan's. The French-Swiss botanist Augustin Pyramus de Candolle kept a mimosa in constant light and discovered its sleep-wake cycle ran shorter than the day-night cycle of twenty-four hours. Since no known earthly force could be triggering this rhythm, he hypothesised that it came from within the plant.

Despite the evidence piling up, doubters argued that some hidden external geophysical cue must be at work. Only in the twentieth century were scientists able to confirm with certainty the existence of an inbuilt biological 'clock'. The modern explanation is very different from theories proposed in the eighteenth century. Nonetheless, de Mairan's investigation is considered the first experiment in chronobiology – the study of biological rhythms in all organisms.

The movement of the sunflower bud, it turns out, is regulated by its internal circadian clock rather than its attraction to the Sun.[47] The rooster's internal clock prompts him to crow before daybreak.[48] And the presence of the internal clock in the human body means that at fairly predictable times of day or night, we grow sleepy or feel hungry, among many other biological processes. Rather like the plant, our circadian clock is regulated by the Sun rising and setting over our horizon. In that sense, we are solar kin.

Yet the internal clocks humans and other mammals possess are structured differently from a plant clock, Alex told me. If you or I were to fly with a plant to another time zone, our own circadian rhythms would take longer to adjust to the local day-night cycle. To put this revelation simply, plant clocks are more agile than ours.

In the botanic garden there are thousands of species of plants from around the world which *don't* have rapid rhythmic movements, unlike the special selection in the Circadian Beds. Most flowering plants, that is, don't move like the morning glory or the marigold. So why do some expend energy on folding and unfolding?

One significant influence is the vital relationship that many plants have with their pollinators. These are the bees, flies, moths or small mammals that carry their pollen where the plants themselves cannot reach and whom they reward with the sustenance they provide. Certain specialist species have evolved to be pollinated by a narrow range of pollinators – and some have flowers that open only when their primary pollinators are most active.

Beside the Circadian Beds, there's a label for a tobacco plant, *Nicotiana sylvestris*, from Bolivia and Northwest Argentina. Its white flowers open in the evening and intensify their heady scent to lure the moths that come out at night in their native habitat. The botanic garden's director, Beverley Glover, later explained to me that when a flower has evolved to attract a particular kind of pollinator, it makes sense that it focuses its effort on their peak foraging times. Otherwise the flower would risk wasting energy on attracting and feeding visitors at other times of day who aren't designed to distribute its pollen – and who might even clog its stigma with pollen from an incompatible species.

Beverley told me, too, about another extraordinary tobacco plant that radically shifts its daily routine because of the trouble it has with its primary pollinator. The *Nicotiana attenuata* is native to Southwest USA. Normally its flowers open at night and release scent to draw certain species of hawk-moth. But while they're visiting, they're liable to lay eggs that hatch hungry larvae. Ironically, the plant comes under attack from the offspring of the very creatures that enable it to procreate. When this happens, the agile plant shifts to produce flowers that open in the morning, when its preferred pollinator is the hummingbird.[49]

Alex and I stood between the day-and-night beds and imagined them flourishing in a few months' time. During the day the many-coloured flowers on our right would be alive with bees carrying pollen tightly stowed on their hind legs and with pollen dusting their furry coats. At night the pale-petalled plants on our left would be glowing in the moonlight and fluttering with moths.

We retreated from the cold to the tropical wetlands house, where the warm air was full of earthy scents. Once my glasses had stopped steaming up, we sought out another plant renowned for its movement: the *Victoria cruziana*, a giant waterlily from tropical South America. We found a little leaf-pad floating alone on a dark pond. It would be one metre wide by July, when its huge buds would blossom and be carefully pollinated by paintbrush by the keepers of the hothouse.

In the wild, the pollination of this giant waterlily entails an especially long and elaborate sequence of flower-insect choreography. The event begins at day's end among the multitudes of white female buds that fleck the surfaces of waterways in the Amazon. First the buds unclasp to release warmth and the scent of pineapple. Dusk-flying scarab beetles are drawn in and gorge on the starchy yellow tissue inside the flowers. At dawn the petals close, trapping their insect visitors, who carry on feeding. Now the white waterlilies transform, Orlando-like,[50] into male flowers with pink petals. At sunset the flowers reopen, releasing the beetles, who emerge covered with pollen. They fly off to repeat the cycle with the evening's new batch of white water lilies. The pink lilies, now pollinated, slip beneath the surface of the water. On and on goes this intricate rhythm of reciprocity between pollinator and pollinated.[51]

Alex and Beverley and the team at Cambridge University Botanic Garden didn't attempt to plant flowers in hour divisions to perform as a timepiece, given how notoriously difficult the task is reputed to be, even for specialists.[52] But they do know someone who has. Their former colleague, Sylvain Aubry, is one of the doughty few in recent years to have experimented with putting Linnaeus's scheme into rigorous action. In 2022 he led a team to grow the floral clock at the Botanical Garden in Bern in Switzerland.

Sylvain is a biologist who researches how plants evolve to cope with variations in day length. But like Alex, he wasn't aiming to prove anything about how green things tick. These days the cutting edge of research on plant chronobiology is at the molecular level. Rather, the Bern experiment was 'vintage science': a temporary installation inviting visitors to explore how plants move in response to their surroundings.

'It was really intense labour,' Sylvain said, when we spoke a couple of years later. The Bern floral clock had its own dedicated gardener and 'it took serious work to keep it going from May to September'. Before they began, the team had to figure out what to grow, given the species on Linnaeus's list don't all flower at the same time of year, nor keep the same daily timings in Bern as in Uppsala. One advantage they had is that there's a greater array of commercially available plants to choose from today. Yet different varieties of courgette, say, produce flowers with divergent routines. So, the team tested more than fifty candidates over two summers. 'We ended up,' Sylvain said, 'with twenty-ish species that would make a decent Linnaean clock' in the garden at Bern. Depending on the month, the passing hours would be measured by morning glory,* trumpet gentian,* goat's beard, common rockrose,* chicory, cat's-ear,* eleven-o'clock-lady,* *Gazania*, night-flowering catchfly.*

And did the experiment work? 'Most of the time,' Sylvain confirmed. 'Gazanias are perfect clock plants, if it's not too cloudy': their opening and closing times didn't vary by more than half an hour over the whole summer. The gentians were 'very good,' he added, 'but very sensitive to rain.' As for individual flowers, like us all, how much they move depends on age or exhaustion.

* *Ipomoea purpurea.*
* *Gentiana acaulis.*
* *Helianthemum nummularium.*
* *Hypochaeris radicata.*
* *Ornithogalum umbellatum*, commonly known as Star-of-Bethlehem.
* *Silene noctiflora.*

There are many different strategies going on among these species that are millions of years apart in evolutionary terms, Sylvain said. And it isn't always entirely clear why particular species fold or unfold. Some are synching with their pollinators. Others may be closing to protect their delicate reproductive centre from daytime heat or nocturnal chill – and in some the internal clock may be more in charge than in others. While the team expected to find certain varieties of flower closing early if they'd been pollinated, they were unable to confirm if that was happening with certainty.[53] But as far as they could observe, they found some species in Linnaeus's floral clock table, like chicory and goat's beard, quite significantly altering their opening hours. That change in timing, Sylvain reflected, was probably largely influenced by a combination of fluctuating day length and temperature.[54]

Marshalling flowers to act like a clockwork machine turned out to be even harder than this experienced team had imagined. But that's not because plant clocks are blunt and basic. After speaking with Sylvain, Alex and Beverley, the more I understood the opposite is true. Plants have an astonishingly sophisticated ability to sense and keep time in response to the world around them.

The promise we might perceive in Linnaeus's idea is a spectacle of regimented flowers keeping identical clocklike hours. But what kind of time does a timepiece made from living flowers actually give?

Sylvain told me about his conversations with visitors and colleagues around the clock growing in the soil and sunlight at Bern. One of their major insights was how the floral clock doesn't represent our kind of time but *flower* time and *pollinator* time. This was a 'timepiece' composed from plants whose own rhythms are shaped by the light, the warmth in the air, the touch of a bee. And that sensitivity to the world, especially the day length, 'put into perspective the very rigid time Western clocks keep', Sylvain said. 'We feel it ourselves when the nights are drawing in. We experience how our standard clock is mismatched with the world.'

As for my own experience of the animated flowers scattered through the streets and parks around me, their rhythms feel so contrary to the mechanical beat of the city. There is a gentle but almost radical pleasure, I think, in sensing the time from the motion of their petals and leaves.

One evening I was talking with Jmeel and Shab about the sky-blue chicory and the purple shamrock. And they put into words what I find most captivating about them. 'Their movements,' Jmeel said, 'remind us plants are truly alive and not just a decorative backdrop to our human drama.' All plants, that is, not just those whose unfurling or tilting captivates our attention.

My mild obsession with flowers with rhythmic movements, I admit, might sometimes lead me to over-identify. I perk up when I find a plant that seems to be closing its eyes or napping at noon in an uncannily relatable way. One risk of projecting human traits onto more-than-human life is that we fail to recognise other species' differences. But playfully nodding to a plant whose own head is slowly swivelling could be, I feel, a fondly absurd act that acknowledges both our shared existence and our utter uniqueness.[55] What I do know is that my fascination with all kinds of plants has gradually deepened through devoting time to the habits of a small group. I am much more attentive to the astonishing lives of plants, and aware of the pollinators and people who tend them.

THE FULLNESS OF TIME

After that summer when Jack-go-to-bed-at-noon flourished for a season in our carpark, we never found the plant growing here again, despite me trying to sow some of its seeds. In Jack's place, I planted California poppies* with citrus-bright blooms. That means Shab still sometimes finds me lurking by the bike shed admiring a flower folding itself up. I've sown night-scented stock* in the carpark, too, so its lovely perfume mingles with the jasmine and honeysuckle.

When I wander about peering and sniffing at flowers in the soft evening light, occasionally I hear someone washing up at their kitchen window and singing along as they clatter the plates. It's a pleasant way, I think, to get the job done.

Now we head to a time and place where everyday life was full of people singing as they went about their work.

* *Eschscholzia californica.*
* *Matthiola longipetala.*

3

Waulking Songs and Furlong-Ways

Beating time by the human voice and body

It's a mild autumn morning on a small storm-swept island in the Scottish Outer Hebrides. The sheep are grazing in the rough grass on the hill. Grey gulls float in the calm sea loch. A little boat is moving fast over the water and the rowers are singing to the beat of the oar. Their lilt carries to the field by the shore, where a girl sings to the cow she's milking. She nods to the women carrying the boards they've borrowed from the footbridge back to the croft. When they reach the barn, the women set the boards on stools to make a sturdy table. They're getting ready to finish a new length of tweed by kneading it until it shrinks and softens. They lay out the heavy loop of wet woollen cloth on the table and settle down around it. Now they pound their palms on the boards and, as the beat builds up, they begin to sing. The women sway their bodies like the rowers on the water – gripping, pounding, pushing, passing the cloth loop in synchronised motion – to the quickening pulse of their song.[1]

This musical, lyrical, rhythmic state is how it might have been on an ordinary day among the fishing and crofting communities of the Outer Hebrides three or four generations ago. In the West now, not many of us associate creating music with work. But this is a place where song and poetry were tightly woven into everyday life – much of it spent making things, fishing, cultivating, tending, and performing hard physical graft. The Gaelic tradition-bearer Annie Johnston (Anna Aonghais Chaluim), who was born in 1886, wrote that a 'large part of' the songs from her home Isle of Barra 'were labour songs – songs of spinning, weaving, milking, herding, rowing, songs which lightened labour, and made communal work of every kind a pleasure rather than a burden'.[2]

There are song traditions from all around the world that lift the spirits and relieve tedium while performing a long, heavy, repetitive task.[3] And there are certain kinds of song that help the work go better by driving the pulse and pace of an action or coordinating bodies to move in harmony together. More than that, to adopt a useful distinction, certain labour songs serve both to *keep* time and to *pass* time.[4]

Many of the songs and chants in this chapter that were used to keep time – to synchronise movement, manage tempo, measure duration – are less precise but capable of more than the clock, metronome or time-tracking app. Reciting a prayer might bless the work. A song may be tailored to the rhythm of the particular task. And as Annie Johnston observed, singing could relieve the burden.

One especially beautiful and intricate example of singing while performing a collective task is happening in the opening scene, above. The women are waulking (fulling) the tweed, like the group in this photograph, below, taken on the tiny Hebridean island of Eriskay in 1934.

The process of making the tweed cloth was a communal effort that began in the summer when the coarse wool was

WAULKING SONGS AND FURLONG-WAYS

clipped from the island sheep. The wool was washed, teased and carded (combed), then spun into yarn and woven at a home loom. The wool would be dyed, perhaps with nettle (grey-green) or heather tops (yellow) or other substances for other hues. The colour was fixed by soaking the cloth in stale urine. Finally, usually in the quieter months of the crofting year, the waulking (*luadh*) was performed by beating the cloth, potentially for hours.

The purpose of the waulking was to shrink and bind the textile and give it the softer, thicker, more airtight finish that would protect the wearer in the wild weather of the North Atlantic. The islanders had their own local ways of performing this heavy work, and there were no hard and fast rules, but it was always done while singing. And of all the traditional Hebridean songs that survive, the waulking songs (*òrain-luaidh*) are the most abundant.

Let's return to that songful scene, as it might have been, in a barn or weaving shed at one of the women's homes.

THE FULLNESS OF TIME

Over the drumming of the hands on the board, the soloist sings, 'How wretched that just for a week I cannot be a wild goose or a grey gull.' While she draws breath, the others answer with a chorus of syllables with a strong pulse: 'O ho rò hò gù.' The singer goes on, 'I would swim the great ocean', and the women repeat, 'O ho rò hò gù' ... 'I would scale the castle' ... 'O ho rò hò gù' ... 'I would crack the locks' ... 'O ho rò hò gù' ... 'I would break you free' ... 'O ho rò hò gù.'[5]

The wet woollen loop is scratchy and stinks. The waulkers' hands are cold and raw. They're breathing hard from working the cloth. Yet on and on they sing the 'beating songs' (*òrain-bhualaidh*). When their bodies are warm and in the swing of the work, the leader of the group gradually quickens the pace. As the songs go on, the women move faster and faster, never pausing for long, so as to avoid bringing bad luck to the cloth.[6]

The waulkers sing of the sea and land: birds and bladderwrack, deer and fairies. There are stories about milking, herding and hunting. There are old ballads about the clan and chief, and battles lost and won. There are harrowing laments grieving drowning, famine, destitution, violation. In one song, a mother fears her child will fall ill with croup during a potato blight.[7] More than once, we hear the voices of women abandoned by a 'well-born' lover: 'This year I am no more to you than the birds of the air.'[8] And there are loving, lusty lyrics: 'You're my hazel-nuts, my berries, / You're my grass nests of wild honey.'[9] These various verses might have been uttered first by women waulking last winter, or a generation past or four hundred years ago – and each singer now gives her own meaning to the songs.[10]

As the waulking goes on and the wool becomes easier to work, the beating songs grow lighter in spirit, and their forceful pulse accelerates. After perhaps seven or eight songs, the leader measures the width of the cloth with her hand to see if it has shrunk enough, and decides it needs just two more songs. When at last the thickness and texture is right, the beating songs finish and the waulkers roll up the cloth to the slow 'folding song' (*òran-pasgaidh*). Once the cloth is rolled, the energy shifts again with the 'clapping songs' (*òrain-bhasaidh*). These quick-tempo songs are sung while two women face each other and beat the cloth hard with their palms, at perhaps two beats a second.[11] The clapping songs are fast, funny and mischievous. 'I won't marry an *old* man,' they sing. 'He'll think the Moon is the Sun. He'll think the snow is sugar. He'll think the rocks are sheep. No, no, no. I shall marry me a *young* man who will lift me up and lay me down!'[12]

Now the waulking that began with the slow sad songs sends the cloth into the world in merry spirit. The waulkers celebrate with tea and cakes, and more singing, dancing and jubilation.

There can be very few people today who have waulked tweed by hand and song as many times as Frances Dunlop. She's the leader of Sgioba Luaidh Inbhirchluaidh, the singing group she founded some two decades ago in the district of Inverclyde (Inbhir Chluaidh) west of Glasgow.[13] Sgioba Luaidh specialises in the Scottish Gaelic (Gàidhlig) labour songs, particularly the waulking songs. These days they're sometimes performed without the drumming of hand on board. But for Sgioba Luaidh, the sound and motion of the work is integral to the music. 'On a good solid wooden table, you get this thump, thump, thump … And it really does add to it.' We're speaking on a video call one evening in 2024 and, as Frances beats down with her hand, her room quakes violently on my screen. 'See! We used to practise in my house, and my table is not what it used to be.'

Most often, Sgioba Luaidh sing the songs while beating a *dry* loop of cloth. But the group is unusual in having properly waulked a length of hand-woven tweed more than a dozen times. The cloth, though, was soaked in a hygienic laundry substance and, thankfully, never in stale pee. ('Health and safety nowadays,' Frances says, 'you couldn't use the real thing.')

I'm keen to learn from Frances why, in her experience, the waulkers would throw their energy into singing while they're performing heavy labour. 'Well, you should see them when they're not singing – they're all over the place! The rhythm of the song helps you to keep the beat. And you're passing the cloth round the table all the time so that it's worked evenly.' Singing lifts the spirits and helps 'the work to go along, because it's very tiring. You could be thumping away for two or three hours or more.'

The waulking was women's work, and it was very carefully performed. 'It had to be – a man's life could depend on good, strong weatherproofed clothing,' Frances explains. 'There's a Gaelic song that says, "It wasn't the cold that took off my old man. No, it was cloth not properly waulked."'[14] That's how important the waulking was. 'They took great, great pride in this work of finishing the cloth.'

The chorus is the part of the song 'that gets the work done', Frances says. This is the group refrain that typically follows each verse sung by the leader. The chorus is made of rhythmic syllables called 'vocables' – like 'O ho rò hò gù' – which are unique to each song. The vocables aren't words as such but sound like an answer from the group to the soloist. The chorus is sung with a strong driving beat to the rhythm of the hands thumping the board. One witness wrote of how '[t]he movements of the women, at first slow, are in perfect rhythm, and, like all co-ordinated movement in these islands, their direction is *deiseil* – sunwards' (the cloth is turned 'clockwise', in accord with the Sun's apparent path).[15]

The leader typically guides the waulkers like the cox of a boat, singing the verses and driving the pace as they work the cloth. Each song may have forty or fifty verses, so while the waulkers sing the chorus, she has the chance to catch her breath and recall the next line.

If you're among the waulkers, how does it feel to keep repeating the action and chorus over a very long, accelerating cycle? I put this question to the leading Gaelic singer, Fiona J. Mackenzie, who shares with me her own experience of waulking tweed by hand. 'You're very aware of everybody else,' she says. 'You're physically close to the people either side of you and you're moving as one body. It's exhausting and you're not thinking consciously about how to do it, you're just getting on with it.'

Both Frances and Fiona describe the feeling that builds as 'hypnotic'. For Annie Johnston, writing in the early to mid-twentieth century, that sensation is a key to how the songs made work 'a pleasure rather than a burden'. It was the 'hypnotic effect' of the tune and rhythm, she wrote, that kept the waulkers going 'till almost exhausted'.[16]

In the 1950s the Gaelic song expert Margaret Fay Shaw recalled that the person singing the verse line sang 'always in absolute time and with a rhythm that was marvellous to me'.[17] I had assumed that meant the leader kept rigidly to the beat. But as Fiona explains, 'There'll be missed beats, there'll be skipped beats, there'll be emphasised beats that are before the bar line.' In this oral tradition, the word takes priority.

'You can always tell a really true traditional singer if they're singing the words as if they were speaking them.'

If you've been brought up in the Western classical or pop tradition, this is hard to grasp. Even Frances took a year to unlearn her classical training, where you 'sing the time values, you sing the dotted crotchets'. One of the major challenges in keeping hold of the old songs, she tells me, is that when they're written down in modern notation 'it sets them in stone, it sets them in concrete, and you haven't got the fluidity, the flexibility, really, that they should have'.

Notation gives a standardised, metronomic tempo to rhythms that are more elastic in the flesh. Fiona agrees, having tried to transcribe the songs many times and ending up abandoning bar lines in favour of 'one long stave'. But if you're singing with an instrumental band, she reflects, 'You *have* to sing within the bar lines, and that does change the character of the beast.' As Fiona and Frances both express, this is a worry for the future.

I'm not a musician and only sing to lighten the tedium on long car journeys (by which I mean, wail along to pop songs with a typical time signature of four beats a bar and a tempo of around 120 beats per minute[18]). To help me understand better what this all means, Frances sings me a waulking song. It takes me a little time to set aside my expectations, but then I begin to hear it. While her hand drums regular beats, the words run around them like waves.

One of the most vivid accounts of how life used to be in the Outer Hebrides comes from Margaret Fay Shaw. In 1929 the young American musician travelled to the far edge of the North Atlantic to hear the Gaelic people's 'own everyday songs, songs that they sang when they rocked the cradle or worked the spinning wheel'.[19] For the next six years she lived on the long Isle of South Uist with two sisters, Peigi and Màiri MacRae, in one of the thatched cottages that stood, she wrote, 'like haystacks on the rim of the world'.[20] There were terrible gales and no electricity. Yet she recalled that it 'was a life that I loved'. Shaw devoted herself to learning the language and culture from the fisher-crofters. 'I loved the songs,' she wrote. 'The tunes were heavenly.'[21] She loved the islanders, too, especially the MacRaes.

Shaw took this fine photograph, above, of Peigi MacRae with Dora the cow on South Uist in around 1930. Dora would not give milk without a song,[22] and so Peigi must be singing in this picture, in time with the steady rhythm of her milking hands. Shaw describes Peigi singing while doing other tasks, too, as if the songs flowed from the work. When Peigi was at the spinning wheel, she wrote, the 'rhythm of her foot on the treadle brought forth the songs as naturally as her fingers turned the wool to yarn'.[23]

There was a time, Fiona explains to me, when the Hebridean people sang while doing any kind of task. 'Singing was their life. It was the rhythm to their life. Whether it was sweeping the floor or feeding the chickens, everything had a rhythm to it. Whether they were hanging the washing out or beating the sheets, there was always a song. Whatever they were doing had an action and a melody to go with it. It was just what you did.'

The songs were tailored to the rhythm and speed of the work in hand. Different parts of the waulking call for the beating songs, with their quickening pulse, or the slow folding songs. On Uist and Barra, there were the rapid clapping songs. And there are other songs that carry the tempo of other tasks. As an example, Frances sings me a spinning song, 'Fuirich An-Diugh Gus A-Màirieach'. It has a bobbing, accelerating rhythm that follows the wheel round and round, then pauses on one long note. 'That's the *draft* of the yarn,' she says, where the spinner's arm extends to draw out the thread. Sadly, I cannot grasp the words because I have no Gaelic. Still, the unexpected rhythm and rolling melody are entrancing, and Frances sings with such feeling in her gentle voice.

Querning songs follow a different movement. The quern is a very ancient tool used around the world for hand-milling grain. It's made of two stone discs, one on top of the other, with a well cut into the upper stone. The grain is poured through the well, and the upper stone is turned round and round to grind the grain against the lower stone. The rotation can't be done in a steady motion, so the songs that go with it have an uneven rhythm.[24]

Querning died out in the islands many generations ago. In the eighteenth century, landlords had the quern stones broken and forced tenants to pay for their grain to be ground mechanically.[25] One of the few querning songs to have survived today is a light-hearted tune called 'Brà brà bleith'. In a recording of Annie Johnston singing this song in 1950,[26] you can hear how the refrain has a very regular tempo. But, in between, the rhythm of the words is stretchy and complicated.

Since the Gaelic songs are an oral tradition, they may change as they pass from one voice to another, and one place to another. Frances talks me through the lyrics she knows to this song. Old woman, goes the verse, grind the quern and you'll get bannock bread. It's not my job, the woman answers. Grind the quern and you'll get cheese. It's not my job, she repeats. Grind the quern and you'll get the son of the man of the house. *Ooh*, she says, *I'll grind! I'll grind!* Frances chuckles at the woman's glee.

Singing while working endured in the poetic culture of the Hebrides until perhaps a lifetime ago, with the tweed continuing to be fulled by hand and song in some parts of the islands until the 1950s. And though we might struggle to imagine it now, at various times and places, singing was a part of everyday work right across Britain.[27]

Picture the Pennines in northern England two centuries ago, especially the Dales between Richmond and Kendal. There are grassy fells crisscrossed with sheep runs, footpaths, pony tracks and tumbling becks. There are moors and marshes where the curlew calls and the snipe drums. There are pastures where the lapwing lays her eggs in the grass. Down where the fellsides roll in soft scoops to the valley floor, there are stone-walled farms and villages among the sheltering trees. And all around you – in the sheepfold, on the moor, in the lane, by the stove – there are people knitting while they go about other kinds of work.[28]

When the artist George Walker arrived in Wensleydale in the early nineteenth century, he found everyone knitting whenever they didn't need to use their hands to do something else. 'Young and old, male and female, are all adepts in this art,' he observed. 'Shepherds attending their flocks, men driving cattle, women going to market, are all thus industriously and doubly employed.'[29] Other visitors recorded similar scenes going on in the eighteenth and nineteenth centuries in this remote Dales region of Yorkshire and Westmorland (now Cumbria). For the people here, making woollen stockings, gloves, hats and so on for the town markets had become a way to top up a meagre living gained from – and evidently while performing – other kinds of labouring.[30]

Aptly for a shepherding community, the Dales knitters counted stitches (loops) by chanting the number-words used for counting sheep: *Yahn, tayhn, tether, mether, mimph*…[31] And they sang counting songs that helped the knitter to pass the time, keep a regular action and work fast.[32] Dentdale was especially renowned for its knitters. There is an old poem that says a 'clever lass' from Dent could knit and sing and carry her milking pail and drive her cow to pasture all at the same time.[33] If this scene wasn't too much of an exaggeration, maybe singing would help such a champion multi-tasker to coordinate her knitting hands and walking steps as well as encourage the cow.

The Dales knitters are remembered for their lightning pace, aided by a special technique using curved needles. In *The Old Hand-Knitters of the Dales* (1969), the historians

Marie Hartley and Joan Ingilby explained the 'secret of the method is the rhythmic up-and-down movements of the arms'. The knitter's body sways along 'with this action which is something like the beating of a drum', they wrote. 'It is impossible' to knit this way 'in slow motion and the loops fly off quicker than the eye can see'.[34]

Every member of the household knitted, from the elderly to the very young. In other words, this never-ending, unbounded form of industry relied on child labour – and the experience could be very bleak. One elderly woman looked back to her time as a small child at a knitting school in about 1760. The children knitted as hard as they could, while singing a repetitive song. If they hadn't done the day's work set for them, they weren't to go to their dinner. 'Neet an' day,' she recalled, 'ther was nought but *this* knitting!'[35]

If her school was like the lace schools in central and southern England, it would have been truly grim. In those places, children as young as five were exploited to make bobbin lace. Work discipline was controlled by cane and 'proficiency' was 'estimated by the number of pins placed in an hour', according to one authority.[36] At a lace school in Northamptonshire, for example, 'the girls had to stick ten pins in a minute, or 600 in an hour; and if at the end of the day they were five pins short, they had to work for another hour'.[37] To drive the pace and keep track of 'the amount of work to be got over',[38] lacemakers chanted counting songs called 'tells'. In the schools, after each count was called, there was an enforced interval of silent work called a 'glum' (a time of gloom). If a child looked up or spoke too soon, the glum would be repeated.[39] As scholars discern today, the lace school tells are an especially clear case that reveals the cruel potential to songs which drive speed, focus and regularity.[40]

THE FULLNESS OF TIME

One crucial distinction between ways of singing while performing a physical task, the historian Ken Mondschein suggested to me, may lie in the 'embodied experience' of the person working. That is, whether they feel the timing is forced on them and conflicts with their own rhythm and speed, or flows naturally from the task and draws them into the 'communal body'.

Something like that last experience does appear to have been part of life in the Dales. There is an account – a warm memory – from the Cambridge don Adam Sedgwick, who was born in 1785 and grew up in Dentdale. He recalled neighbouring families coming together in the evening for sociable 'sittings', where they rocked away at their knitting together beside the fire. While working 'with a speed that cheated the eye', he said, they told stories and sang 'ancient songs of enormous length'. At these companionable sittings, Sedgwick wrote, 'labour and sorrow were divorced, and labour and joy were for a while united'.[41]

One of the few Dales knitting songs to survive[42] is a rhythmic ditty from Dent about Rockie the dog running around a local hill after lost sheep (a panicky, heart-racing drama familiar to any shepherding community). During each verse, the knitters knit one more round of the stocking, until they've knitted twenty rounds. As if to propel the knitters along with greater speed, each verse ends with a rollicking, 'Run Rockie, run Rockie, run, run, run.'[43] There's an obvious giggle to be had in calling out together to urge Rockie to chase after the errant sheep while you race to knit a round.

But by the mid-twentieth century, Hartley and Ingilby found, memories of this very old industry had nearly disappeared, and there were 'no men and few women' in Dentdale who could still 'knit in the old way'. The beautiful gloves that had been made for private buyers earlier in the century were 'a last flowering of the art of the old knitters', who had been born into a life where 'skill in the craft' was passed down through many generations.[44]

My grandfather grew up in the Dales in the early twentieth century not so very far from Dentdale. My mother had a fond pride in his ability to knit. It was a remarkable thing for a man to do in our community further south, where knitting was thought of as exclusively women's work. I have always put it down to his amazing ability to create things and wanting to keep his hands agile. But are those the only reasons? After reading about the Dales knitters, I wonder now if he knitted partly to hold on to the presence of people long lost to him, by making as they did (though not in the old style). Yet there is no one left for us to ask. The memory – the yarn that connects us – is becoming harder to hold.

The famously rapid craft of the Dales knitters leant itself to a local expression for a brief duration. In the poem *Reeth Bartle Fair* (1870) by John Harland, a Yorkshire lead-miner is walking back from the fair when he bumps into his friend Curly:

'Swat te down, mun, sex needles,' said he,
'An' tell us what seets te saw there.'

Curly wants him to sit (swat) down and tell him about the sights he saw for the length of *six needles*, literally the time the knitter takes to 'work the loops off six needles'.[45] This would take me an age. But in the Dales 'six needles' was an everyday way of saying 'a short while'.[46]

That loose analogy is particular to a region with a specialist craft. Yet across society, people did once measure spans of time by common techniques using rhythms of the voice and body. There are recipes, for instance, for boiling eggs or mixing paint that refer to 'a Pater Noster while'.[47] That's the time it takes to recite the Lord's Prayer. More than just a utility, reciting a prayer was a pious act and it's easy to imagine it was understood as a way to protect the matter in hand from evil influences.[48]

For a greater span, you could repeat the prayer or recite a longer one. Or you might say it more slowly – as in a recipe recorded in the seventeenth century by Sir Kenelm Digby, which advises against over-stewing tea leaves: '[t]he water is to remain upon it, no longer that whiles you can say the *Miserere* Psalm very leisurely'.[49]

Then there are the involuntary rhythms of the body, with hunger being potentially the most pressing sign of the time. And this has been tested, to an extent. In the mid-twentieth century, researchers isolated male graduate students 'in a dark, lightproof, soundproof cubicle' for durations ranging from eight to ninety-six hours. On release, each participant was required to guess the day and time (and in general their answers fell well short). Later, each was asked what 'cues he had used in arriving at his estimate'. Most relied on 'observable bodily processes', with 'eating and hunger' then 'beard growth' at the top of the list, followed by 'urination, defecation, and sleepiness'.[50]

In the real world, how consciously and explicitly would people have marked time not only by their rumbling stomach but by other biological processes? According to a source quoted by the historian Nina Gockerell, there's an account from the late eighteenth or early nineteenth century in Styria, Austria, of a case where a suspected thief gave the time of the crime as '[s]omewhere around the second visit' – meaning the second nocturnal time for urination (3 a.m., the first being 11 p.m.).[51] I am sceptical about how clocklike trips to pee would be in actuality. But they would be a way of dividing up a long night into different phases.

If you'll forgive me sticking to the theme for a moment, one easily understood expression for a short duration is a 'pissing-while'. This attention-grabbing phrase crops up in comedy plays of the sixteenth and seventeenth centuries.[52] A sedate version is employed, for example, by the Widow Blackacre in *The Plain Dealer* (1676) by William Wycherley: 'stay but a making-water while, (as one may say) and I'll be with you again'. Was this an everyday equivalent to 'just a jiffy'? Or mostly just a running joke in the theatres?

A less larky and potentially more definite measure is the time it takes to walk somewhere. In the late fourteenth century, Chaucer refers to a short while in his verse as 'a furlong way'.[53] That's the time it takes to walk the length of a plough-furrow, which is reckoned to be around two or three minutes. And it might have been a common expression. I've noticed the 'forlongwey' in a fifteenth-century recipe for Fried Cream of Almonds.[54]

Longer spans of time may of course be measured by longer walks. A sixteenth-century remedy for toothache directs the sufferer to hold the ball of medicine between the cheek and tooth by 'the space' that 'one may go a mile'.[55] Luckily for the patient with the throbbing jaw, probably that doesn't mean marching about was part of the remedy. A furlongway or a mileway are durations people could likely guess from experience without leaving their chair – just as we can estimate when two or three or twenty minutes have passed.

A 'mileway' could imply a different span of time, depending on the season.[56] That makes sense: the track to the next village, say, would be slower going in mud or snow. Chaucer, however, defined the mileway as a fixed duration. In the guide he compiled on how to use an astrolabe (a sophisticated astronomical instrument for tracking time by the Sun and stars, among other tasks), he explains that '3 mile-wei maken an houre'.[57] An hour is the time it takes to walk three miles – although that may have meant a longer distance for Chaucer than it does for us.[58]

From our point of view, it may seem unlikely that authorities who have use of a clock would use the mileway in an attempt to enforce time discipline at work. Yet this they do in an ordinance of 1370 for Chaucer's contemporaries, the masons working on the York Minster, the city's splendid cathedral.

The order tells the masons to shut up shop on holy days at high noon 'smytyn by ye clocke'.[59] So presumably they can hear its hour chimes. Yet on ordinary *summer* workdays, which are very long indeed, the masons are ordered to start toiling at sunrise and not finish 'untill itte be namare space yan tyme of a mileway' before sunset.[60]

There may be multiple motivations to combine the clock with older methods of reckoning time. In medieval urban Europe, the collective day was shaped by a mix of local signs from the Sun and particular church and town bells, which fluctuated in timing over the year.[61] When the sound of the clock chime began spreading through the fourteenth-century streets, at first it seems to have been absorbed as an addition to the latticework of urban time signals.[62]

Then there are the particular circumstances of the York Minster masons, who would need good lighting to perform very fine work. In a 'world lit only by fire',[63] time would run out for the artful maker when the Sun goes down and begin again when it rises. That limit is reinforced in their ordinance with a wonderful phrase. In *winter* the masons are told to begin 'als erly als yai may see skilfully by day lyghte' to work and stop when they can't. For this keen-eyed craft, the light must be bright enough to *see skilfully*.

The Sun, in other words, defines the start and end of the York Minster masons' day in winter, too – and its times of rising and setting vary, of course. From what I can gather, the clock the masons could hear was most likely a simple mechanism that only struck on the hour. Obviously, too, the clock is not a device for testing the level of illumination. For employers wanting to squeeze as much good work as they could out of the masons on short winter days, I imagine it would make sense to try to extend their day to the very edges of the darkness – rather than limiting it by the hour chimes that come closest to those moments in the light.

For so many of us now, the clock has developed into an all-purpose precision instrument for managing every aspect of life in fine detail. But perhaps the first people to hear its hour chime in the streets considered it to have a narrow range of uses – just as the masons wouldn't have selected the same chisel to cut rough blocks for a column and carve the delicate curls of a leaf. That might explain why the ordinance combines the clock with more subjective techniques of reckoning time by the body and senses.

What stands out most for me among those methods, though, is that extra element to the rule for summer which tells the masons to keep working until a *mileway before sunset*. To estimate the duration of a mileway takes a certain amount of bodily nous. To anticipate when it's a mileway before the Sun goes down, you'd need a particularly sensitive eye for the warm glow on a stone wall or the tilt of sunbeams streaming through windows. The rule sounds tricky to apply even if you have the honed visual sense of a cathedral stonecutter. And far harder still when sunset is masked by 'wete ploungy cloudes', to borrow a fine phrase from Chaucer.[64]

How extraordinary the method seems now, and how difficult to perform. Maybe it worked well for the York Minster masons. But subjective measures like this could be contested in disputes about when the workday ends. And more widely, the effort to avoid that kind of conflict appears to have been one of the forces driving the gradual adoption of the clock.[65]

'What other times of the day have you got in your book? Breakfast, second breakfast, elevenses…?' My friend Sabih sent me this text one morning when my mind was wandering toward the biscuit tin. I laughed then thought for a moment. His message was a playful aside but reminded me of something I'd been neglecting. We do still talk about time by a variety of measures, and not only in the abstract, quantitative language of hours and minutes.

We partition the day in terms of meals: 'it happened around breakfast time' or 'see you at lunchtime'. We say an event happened 'in the blink of an eye', 'in a flash' or 'at snail's pace'. We say, 'I'll be with you in half a pint'. We name times of day by their sensory qualities: 'the dead of night' and 'daybreak'. Even in my urban neighbourhood, there's a sign by a park that says the gates will be closed at 'dusk'.

What's more, we have ways of measuring duration or timing events very tightly by the body and voice. Musicians, of course, are highly skilled at counting time silently or with the tap of a foot. And in the UK during the Covid-19 pandemic, we all became newly familiar with how to measure about twenty seconds without a clock while thoroughly washing our hands to avoid infection. We sing two rounds of 'Happy Birthday'.

Yet consider how the super-precise measure of the modern clock has become enmeshed with daily life – especially in workplaces where productivity is rated in minutes. The history of time discipline at work is long and deeply complex. But it's worth reflecting on one of the most significant ways that time regulation has been taken to an alienating extreme in our own age through the widespread influence of Taylorism.

In the late-nineteenth century, the American engineer Frederick Winslow Taylor proposed timing manual workers as they performed each 'elementary' motion of a process with a stopwatch to identify the fastest technique. Then to use that data to '[e]liminate all false movements, slow movements and useless movements'.[66]

'Under the influence of Taylor,' Ted Gioia observes in *Work Songs* (2006), 'the benefits of harmony and rhythm in the work environment, forces initially arising spontaneously, would now be studied, systematized, and propagated.' What's more, the 'worker would now mimic the processes of machinery, adopting the repetitive motions and unwavering pace of a piece of equipment. The rhythm of manual work would thus be forced to match the tempo and instrumentation of automation.'[67] In the factory, this implies, the organisation of bodily rhythm, synchrony and speed – once issuing from the task itself and perhaps orchestrated by song – would be driven by stopwatch and conveyor belt.[68] Timing would be defined by machine, and the worker would become more machine-like. We're all familiar now with the legacy of Taylorism: the 'scientific management' of workplace efficiency, the standardisation and monitoring of tasks against the clock, and the reduction of skill and creativity to simple repetitive, monotonous tasks.

In the mid-twentieth century the conditions of work in the cities did not go unnoticed by those endeavouring to preserve the traditions of the Hebrides that were fast falling away. In the 1960s the folklorist John Lorne Campbell (husband of Margaret Fay Shaw) looked back at waulking and other ways of labouring by song in Gaelic-speaking Islands and Highlands. Work, he wrote, 'was performed as a joyful social creative activity' in great contrast to the 'monotony of modern urban factory existence'.[69]

A decade earlier, in May 1956, Shaw had presented a radio documentary on the songs of South Uist for the BBC. No doubt islanders like her friend Peigi MacRae still sang while spinning, knitting or rocking a cradle. But waulking the tweed by hand and song was already at its last ebb.[70] Still, Shaw's documentary evoked this disappearing world as a living present. In one of the recordings she played, a group of women with low, soft, cracked voices sing a waulking song, returning to the same sad notes as their rhythm quickens. The microphone is close, drawing the listener into their intimate circle.

Shaw must have been aware of how different this songful life of the Hebrides may have seemed to her listeners in towns and cities.[71] And there is one detail from her radio script that stands out most for me. 'The time needed to finish a waulking,' she said, 'is always calculated by the number of songs needed, and not by the hands of the clock.'[72]

In the early decades of the twentieth century, quite possibly, the fisher-crofters of the Hebrides did have timepieces ticking away inside their homes. I suspect Shaw intends to suggest it was a deliberate choice – even a necessary part of the ritual – that the timing should be orchestrated by song alone. So, I ask

Fiona, Shaw's biographer, for her view.[73] She explains there is a functional reason why the waulkers would not calculate their finishing time by the clock: 'Every cloth behaved differently'. You can't predict how long it will take for a particular piece to have the right shrinkage and texture. Timing by song, rather than a fixed duration, lets you keep adjusting the tempo to the immediate needs of the task in hand.

Moreover, the waulking songs, as we've heard, are more than just a tool for beating a rhythm, synchronising bodies and driving pace. One of their most remarkable qualities is their emotional intimacy: how they hold and share experiences that might have no other outlet.[74] As a whole, Fiona tells me, the waulking songs carry the culture – the language, histories, poetry, beliefs, identity – from person to person and generation to generation. They are a common treasury. The waulking itself was a special occasion that brought people together. It was, she concurs, 'an opportunity to gain support and share joy'. Even today, singing the songs has something of that power.

When Frances's singing group, Sgioba Luaidh Inbhirchluaidh, meets up in a church hall in Greenock on Monday evenings, sometimes 'we're tired and we don't really want to be there', she tells me. 'But by the time you've finished, it lifts you and you feel a lot happier.' (If you've ever been part of a choir, you may have felt the benefits of singing in synchrony.[75]) And then, of course, the Sgioba Luaidh singers aren't just *singing* the waulking songs – they're turning a (dry) loop of cloth around the group. As Frances has said before, '[t]he whole process was – is – very therapeutic. In the songs you can express your every emotion; the hypnotic thumping of the cloth on the table helps to release all your tensions and frustrations; in the company of your friends you can have a good gossip, and talk over problems'.[76] 'That's how it must have been in the old days,' she tells me, 'because there was no counselling in a wee township or a croft.'

Most of all, Sgioba Luaidh hopes to honour the people who shared their stories in these songs as they laboured to finish the cloth. The songs 'dealt with everything in the women's lives', Frances says. To sing their songs, to sorrow and laugh with the women who made them, is 'very special for us'. And to sing their songs while working is 'very special indeed'.

The group has waulked many times at Auchindrain in the hills of Argyll. This was the last farming township in the Highlands to have survived the Clearances,[77] with the final residents leaving around six decades ago.[78] 'We all feel a powerful connection with the past on these occasions,' Frances tells me. 'We are aware of awakening old echoes, and can sense the presence of the ancestors singing with us.'

The past presses on the present in the things that surround us: the dink in the car door, the paw print in cement, the disused tramline in the street, the patched tweed jacket on the hook. True, some physical traces are very faint indeed – in the next chapter, we'll be seeking the most delicate and ignorable lines scratched into stone long ago to trace time by shadows. But even the slightest incision in a wall has a durability that the motion of the hand and the feeling or story that produced it does not.

The old ways of keeping fixed or flexible time – the singing and the chanting, the tapping of the hand, the swinging of the limbs – are fleeting and intangible. The furrow in the field may leave a visible ripple for centuries, but the rhythm and motion of the bodies that worked it are as ephemeral as a gesture. The words that were sung over the furrow while encouraging the ox and reaping the corn are as transient as the voice.

For the Gaelic labour songs that were caught before they were lost, the challenge now is not to lose the rhythms of the voices and the work from which they flowed. That's why Sgioba Luaidh keeps on beating and turning the cloth. We're trying, Frances says, to sing 'the songs as they were meant to be – giving them breath, giving them life'.

4

Scratch Dials and Stick Dials
Dividing the day by the flow of shadows

It's late June 2023 in Woolstone, a small village below the high chalk ridge at the northern edge of the Berkshire Downs, about twenty miles southwest of Oxford. On the steep slopes, the long grass is mellowing to a yellowy cream. In the bosky lanes, the early summer lushness has gone from the leaves. In the gardens, hot-pink hollyhocks and red roses burn bright beneath thatched eaves. A cat slinks past the pub. Nobody else is around except for us as we follow the path to the white stone church.

Rosie and I have travelled from London this morning to walk the Ridgeway, the ancient chalk road that runs east to west along the top of the ridge. One of Rosie's great joys in life is to be swinging her pins on a track through the hills. And ideally, from her point of view, we would already be striding high in the bright expanse of the Downs without detouring through the shady lanes below. But first we're dipping into Woolstone because I've read its church, All Saints, harbours a long-forgotten device for marking times of day, known as a medieval 'scratch sundial'. This is basically a rod or spike stuck into a wall, with lines scraped underneath to mark significant times by its shadow. Although scratch dials are

SCRATCH DIALS AND STICK DIALS

very simply made, their history and meanings are tantalisingly inscrutable.

Many modern sundials are sophisticated instruments designed to find with precision the same 'equal' hours that clocks keep. That is, they dice time into identical portions that never change from day to day or place to place. But my particular enthusiasm is for the marvellously variable rough-and-ready methods past generations had of telling time by shadows: by arranging sticks in the turf, observing the shadow of their own body, or etching wobbly lines into walls.

Scratch dials are not the earliest form of sundial in the British Isles. There may have been dials here under the Roman occupation, which ended around 410 CE.[1] The oldest confirmed examples to survive are found in Ireland and Wales, and were made by early Christian communities perhaps as long ago as about 600 CE. In England, sundials may have been introduced (or reintroduced) a little later.[2]

The Romans divided the time between sunrise and sunset into twelve evenly spaced intervals, often called 'seasonal' hours, which shrink and stretch in span as the days shorten and lengthen. The further you are from the Equator, the bigger the variation. On this luxuriously long midsummer day in Woolstone, each of the daylight hours on a Roman sundial would last about twice as long as they would at midwinter. Only at the spring and autumn equinoxes, when the daylight and darkness are in balance, would the Roman and clock hours have the same length.

Indeed, before the clock chime began to dominate in everyday life in Europe, the major divisions of the day fluctuated between sunrise and sunset through the seasons.[3] In the Middle Ages the hours remained Roman, and originally the times of Christian worship – the canonical hours – during daylight were linked to them.[4] Yet the timing of the canonical hours varied in practice from season to season, place to place, and over the years.[5] And it wasn't only the pattern of the religious day that shifted. The chiming of the bells from the monasteries and churches provided an underlying framework for the secular day for lay communities too. This older society was deeply concerned with the nature and good order of time. Yet far more than I had imagined, the common beat of time did not have a precise and absolute standard, like the units of the modern clock, without which so much about our present world would not function.[6]

One of the rare remaining early medieval English dials, pictured below, belongs to St Gregory's Minster in Kirkdale in North Yorkshire. It was carved about a thousand years ago.

The inscription in Old English around the half-circle declares, 'This is day's sun-marker at every tide' (hour): 'ÞIS IS DÆGES SOLMERCA AET ILCVM TIDE'. The hole at the top would have held a rod protruding horizontally from the wall to make a shadow over the lines below. The crosses on some of the lines strongly suggest they mark times for worship.

To me, the neatly drawn lines look like they would split the day evenly. But their equally spaced divisions wouldn't, in fact, measure equal intervals of time. (To achieve that would require repositioning the lines by a refined mathematical art developed by ancient astronomers.) Although in everyday life, people wouldn't have looked for that kind of precision or noticed the difference in duration between one hour and the next.[7]

Before our trip to Woolstone, I spoke with Ben Jones, a letter carver and sundial designer who runs the British Sundial Society's register of medieval church dials.[8] I asked him for his favourite example in the archive and he chose the Kirkdale dial. Given his own craft, Ben takes sympathetic delight in how the sculptor seems to have run out of space for the lettering on the bottom line and put the remainder hovering awkwardly above it. Despite that, he said, this dial has been skilfully carved from the stone block before being set into the wall in a lofty position. It's too carefully constructed to be called a 'scratch' dial. And it's thought to be too old.

The earliest scratch dials on the walls of English churches are thought to date from around the Norman Conquest in the eleventh century. And they appear to represent a remarkable deterioration in the craft. The scratch dials, you could say, seem to knock the (presumed) straight line of technological progress out of shape.

Among the old ways of telling time by shadows, the scratch dials appeal to me most for their delicacy and slightness, and for how directly some seem to translate immaterial shadow into solid stone. But at this stage I've only read about and seen pictures of them in journals, one of which pointed me to

Woolstone.[9] Faced with my overwhelming eagerness, Rosie has come along in support to see what we can find.

Since the obvious place for a sundial is on a sunny south-facing wall, we wade through the long grass to the back of the church. There's a lovely languor in this hidden corner of the churchyard. Hens are clucking contentedly in the farmyard next door. The sky above is hazy and bright. We cast our eyes over the rough blocks of creamy chalk stone. But the wall seems absent of etchings.

Hunting for scratch dials is a game for keen-eyed, patient souls. The dials are often composed of just a few slender, shallow lines eroded by centuries of exposure to the elements. When you step back, they recede into the general roughness of the stone. To pick them out from their camouflage, you have to scan the wall carefully in sections and stay alert to the slightest sign of a circle or fan.

Once Rosie and I slow down and take that steady approach, we discover a dial waiting for us at the west end of the south wall, pictured below.[10]

SCRATCH DIALS AND STICK DIALS

It's a rough effort: an imprecise fan of lines radiating from a splodge of cement, presumably added much later. Most likely it covers a smaller hole where the gnomon (shadow-maker), in the form of a rod or spike, would have been fixed at a right-angle to the wall.

As the Sun moves east to west through the day, the shadow on the ground of a tree or our own body moves 'clockwise' in the northern hemisphere. But, confusingly to the clock-trained mind, the angle of sunbeams hitting a rod stuck in a *vertical* wall means its shadow moves 'anticlockwise', as the arrow shows on the diagram below.[11]

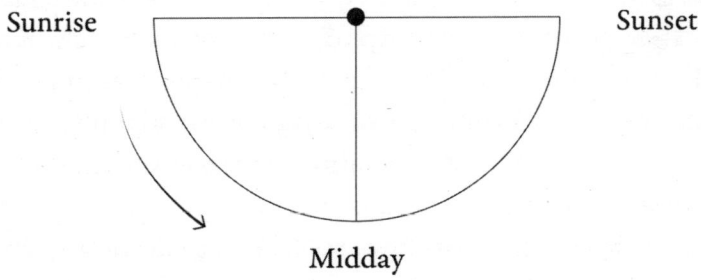

To steady my eye and notice the details in the dial on the wall, I reach for my notebook and draw a picture. As the movement of my hand over the page mimics the hand that cut the lines into the stone, I appreciate just how loose and sketchy they are. This dial doesn't appear to have been made by calculating angles, or transferring a template, or even using something to keep the line straight. There is a theory that some scratch dials were made by finding the time by the position of the Sun, then running a blade under the shadow of the rod.[12] I wonder if the lines on this device, too, might be a fleeting moment in the light experienced in the flesh and scored for ever in the stone.

Strangely, though, the scrappy fan of lines on this Woolstone dial continues above the sunrise-sunset line, where the shadow of a horizontal rod cannot reach. Up close, you can just discern what appears to be a fuller circle of lines for times of night. I'm curious to understand why a sundial would mark hours of darkness.

After the trip I consulted the Italian artist and historian Mario Arnaldi, a person of extraordinarily wide-ranging accomplishment who has created one of the most enjoyable modern sundials I've ever seen.[13] Among historians, Mario is recognised as a foremost authority on medieval European sundials. And he's the person I've learned from most in my quest to discover how ordinary people told the time by shadows.

Medieval sundials, Mario said, were charged with both symbolic and practical meaning.[14] A circular dial that includes the hours of night could be an image of the divine cycle of time as well as a tool for reckoning by shadows.

As for the kind of hours this dial marks, if you count the divisions below the sunrise-sunset line, the day is split into eight portions. Why eight, and not the twelve hours that Christian Europe inherited from the Romans? Because there were all sorts of local practices, Mario explained. In the case of this dial, the eight daylight segments might be a legacy of the old division of the day in England into four 'tides' (hours) and finer divisions of eight or sixteen. And that would presumably explain why there are eight segments on the dial at Kirkdale.

Medieval church dials right across Europe, from Armenia to Ireland, very often share a common shape – a half-circle divided into segments, like the Kirkdale dial. But the number of segments can vary considerably. While it's common for them to split the day into twelve portions, you might well discover dials with four, six or eleven segments. And there are

SCRATCH DIALS AND STICK DIALS

rarer, more inexplicable, varieties that divide the day into ten, thirteen or eighteen parts.[15]

Quite what the lines are designed to mark on this rough Woolstone dial is not clear. Scratch dials, Ben Jones told me, are also known as Mass dials on the understanding they mark the daily times of worship. But he emphasised we don't know for sure if all scratch dials are Mass dials, because most lack the crossed lines (or some other mark) that would suggest times of devotion. Indeed, very little can be said with great certitude about who made them and why. You can't transport yourself into the body and mind of the maker. You can't know what they thought or felt. You can only try to decipher the traces – and that's why they're so compelling.

Yet for many sundial experts, Ben explained, these blunt and mysterious devices lack appeal, not least because they have no mathematics to them.[16] Only in the last century or so have scratch dials become subjects of serious study, and only by a handful of investigators. As a mark of what a niche pursuit this has been, they had been regarded as an 'almost exclusively' English phenomenon.[17] And they do seem to be most abundant in this country. But thanks to Mario and other investigators, we now know scratch dials are to be found across Europe from France to Italy.

For Mario, the difficulty of deciphering *every* form of premodern dial, and what they meant to their makers, presents a splendidly rich seam of never-ending intrigue. There are so many riddles about medieval church dials, he told me, and the scratch dials are the most unfathomable of all.

Astonishingly, for such slight objects made hundreds of years ago, there are believed to be thousands of scratch dials remaining in England. And I have read there is more than one to be discovered here at Woolstone.[18]

Sure enough, Rosie calls out to say she's found another dial, pictured below, by a window near the east end of the church. This is tucked out of sight behind a water butt down an alley. But it is where you'd expect it to be – on the south wall about head or chest height close to the old priest's door, which is now covered by the modern vestry.

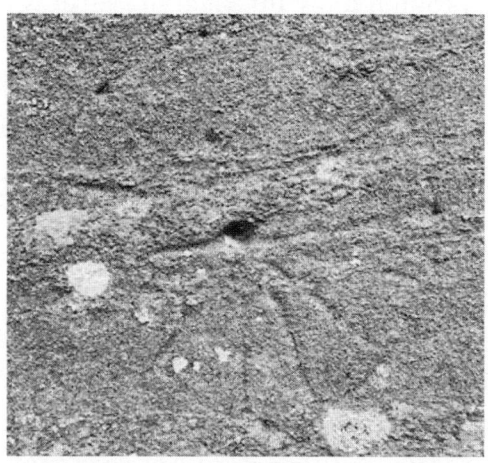

Scratch dials are notoriously difficult to date. But because of the detective work of a past investigator, Frank Poller, we can probably narrow it down for this one. The church is thought to date from 1195. The transept jutting from the south wall just beside us – putting the dial in substantial shade – was built towards the end of the thirteenth century. So, it's reasonable to assume the dial was scored into the stone before the transept and has been here for more than seven centuries.[19]

Even though this circular dial is neater than the first, I wonder how effective its wavering lines could be for telling time. Then I remember Ben explaining that if everyone in the community shared the same means of marking important events, 'you'd all be using the same wonky time and all would be in agreement'.

SCRATCH DIALS AND STICK DIALS

The priest's door is so close to this second dial that I can only guess it was used to tell time for Mass. While I'm drawing it, I imagine a medieval priest in a long cassock poking his head from the door to check the blurry shadow wandering over the dial. Amid the dramas of village life, I wonder whether the agreement about its wonky time ever broke down.[20] Assuming the shadow prompted my imaginary priest to start summoning parishioners, would they ever crowd around him and quibble over the hour? Would they point to another scratch dial? If so, they might have had quite a choice, because there appear to be more beside us on the priest's door.

When Frank Poller visited in the 1990s, he made these sketches, reproduced below, of the marks on the outer stones of the priest's door (now enclosed by the modern vestry). The doorway is covered in a thick layer of white paint, but you can still see several spidery dial-like etchings in the stone. If any of them were actually used to mark time, curiously, they don't seem to divide the day into eight parts like the dials we've already found.

It's not uncommon for churches to have multiple scratch dials. And there are many potential explanations, not all of them satisfying in every case. In some places, maybe the path past the old dial was moved. Or the yew grew over it. Or the new priest simply wanted to make his own mark, quite literally. Or some dials could be 'copycat' graffiti.

When a church has a collection of dials with varying divisions, they might have each been used in different calendar periods for seasonally changing times of worship. Or they may have been made as updates on old models, given the pattern of worship fragmented and shifted over the centuries. Some of those changes were very major, with particular daily times of worship drifting quite considerably from their initial

THE FULLNESS OF TIME

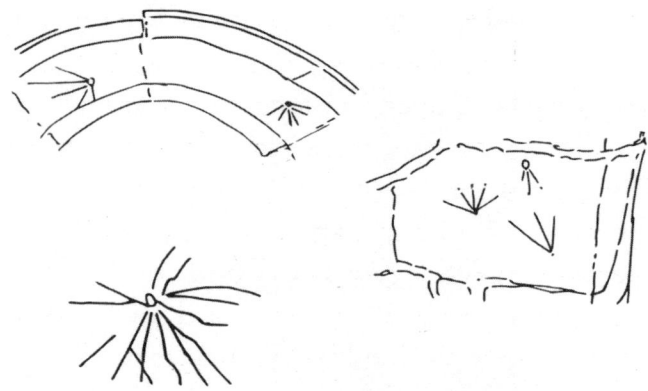

moorings.²¹ One legacy of that slippage lives on in the meaning of the English word noon. 'Noon' derives from 'None', the canonical hour tied to the ninth Roman hour (*nona hora*), which fell around mid-afternoon. But by the thirteenth century, noon could mean midday.²² One old theory is that, in a season of fasting, monks were not meant to eat until None and so its timing was pulled earlier. That might not be what happened, but it's plausible.²³ We might have famished monks to thank for our 'noon'.

At Woolstone, I've read there are more curious marks to be found *inside* the main body of the church. So, Rosie and I wade back through the grass to the front entrance. We push at the heavy door and step inside, where the cool air is vaguely scented with flowers. Near the elaborately modelled lead font, in an anonymous spot on the north wall, we find a deeply gouged fan of five lines tightly squeezed together, photographed below.

The lines stem from a hole which strongly implies it held a rod to make a shadow. If it was a sundial, maybe it marked the time of day through the seasons for a particular event. As for why it's tucked away on the north wall inside the church, sequestered from sunbeams, perhaps it once caught the light

SCRATCH DIALS AND STICK DIALS

from a window that has been blocked up.[24] Or its stone was reused from a former structure in a sunnier position. There is so much to puzzle over.[25]

I leave Rosie pondering the patterns on the extraordinary font and go off to search around the interior door and windows, where the limewash has rubbed away and the stone is exposed. The chalky blocks were cut many generations ago by skilful masons: some are finished with a fine drag, others with a looser rhythm, as if they yielded to the chisel like stiff clay. And over the top of that confident work is a confusion of more tentative marks – layers of little images scraped into the stone with thin tools by less steady hands. A box or diamond with a cross. The letters VV. A ladder, perhaps. A harp – or maybe a shield.

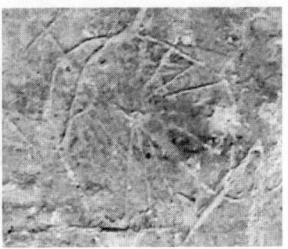

These mysterious sketches are thought to be medieval 'witch marks', as they're often called, to the frustration of experts. That misleading name sounds like they might have something to do with hexing the church. Rather, it's thought they were etched into the body of church buildings as protective charms or 'visual prayers' to repel evil and draw good fortune. Very little is certain about them. Yet it's considered likely they were made by people from all walks of life, often as desperate gestures: to protect newborns from wicked spirits, to resist infection in a time of plague, even to curse a wrongdoer. We could be looking at signs and symbols made by many hands over many years.[26]

Most of the marks are smaller than the dials but have that same shaky quality of line. And some actually resemble the dials. Where does the line fall between these different kinds of mark etched into the stones?[27] After all, sundials would be part of the web of meaning in a holy building where every object has profound significance. I wonder if there was felt to be some kind of blurring of identity here between the dials and the etchings made by many people layering symbol beside symbol, hope upon hope. But all these messages in the walls communicate in ways we no longer fully understand.[28]

After a morning of looking for scratch dials, I'm moved by how little they disguise the endearingly unartful hands of their makers, yet how unwillingly they yield clues about their status and meaning. I'm finding it hard to stop searching for them. And since we're heading now to pick up lunch in the neighbouring village of Bishopstone, I've suggested we make just the tiniest detour for a glimpse of its church.

For such a large village, Bishopstone seems spookily quiet on the day of our visit. At first our only companions are the slow-ambling ducks and skittering insects on the glassy millpond. And we're cheered to hear small voices chirruping from within the primary school as we pass. Soon the sweet sound of the quarter-hour chime locates the Church of St Mary the Virgin among the trees. The path there is rose-scented and seems, to me, full of promise.

The church is grander than at Woolstone and has a much larger expanse of wall. Given we're short on time, I need to make a strategic dash. I've read there's a late Norman door on the north wall that might have been moved from somewhere else. So I take that as my target and jog round to the back of the church.

The stone doorway is indeed wonderful, with a splendidly sculpted zigzag arch. The doorway is equipped on one side with a charmingly clumsy dial whose lines might once have numbered twelve, unlike those we found at Woolstone. I'm pleased to have found a scratch dial so quickly, take a snap and jog back round again to where Rosie is waiting on the sunny side of the church.

She looks preoccupied, restless. Her enthusiasm, I think, is definitely waning.

But coming closer, I catch the light in her eye. 'Look!' she calls. 'This place is festooned with dials.' I'd missed them all in my focused haste.

As many as four sundials adorn the walls of St Mary's church, possibly more, all scored directly into its grey stone. As a collection, they may represent an unusually rich array of eras.

Rosie shows me the two most intriguing designs, which sit on the same buttress in the south wall. The first, pictured below left, is a circle firmly scored with neat hour lines that are placed in a peculiar asymmetrical pattern.[29]

Just above this curious entity is the larger dial, pictured on the right. It has the shape of a medieval half-circle but there's an air of modernity about it. The maker has hacked into the buttress to create a crudely hewn but more properly south-facing surface. The lines cluster more tightly around the middle of the day. The now-absent gnomon would presumably have been tilted to point at the North Star. That innovation, which aligns the gnomon with the Earth's axis, didn't reach England until the sixteenth century, and later still in some parts of the countryside.[30] This all implies the dial was made to measure hours of the same length all year round.

When you peer closely at the rather ignorable buttress, in sum, it harbours two strikingly different ways to divide the day – both of them perhaps too carefully made to be called scratch dials. The lower dial gave some kind of seasonally fluid measure, while the dial just above it would show 'clock' hours. And they might have been used in the same era for various purposes in the life of the church and village.

Rosie leads me along the path to look at another clocklike sundial scored into the stone at the west end of the church.

SCRATCH DIALS AND STICK DIALS

This is high on the tower, on a south-facing surface close to the large mechanical clockface. The lines and numbers are too faint for us to discern from down on the ground. But, uniquely, this sundial is intact, with its rusty wire gnomon bent (presumably) to the North Star. Like all the dials we've found on the walls here, it has been cut straight into the stone tower and not carefully carved by a mason. Still, the time it gave may have been quite precise.

It's easy now to assume the clock made this sundial redundant. In the 1930s, the poet Hilaire Belloc penned a pithy motto for a sundial: 'I am a sundial, and I make a botch / Of what is done far better by a watch.'[31] That sums up what those of us expect who are used to the ornamental garden variety. But Belloc's motto is irksome to those who know that for much of their history, clocks were regulated by the shadow on a sundial or a midday mark.

There's a special object inside the church that might provide a clue about when the clock first struck in this village. Inside the nave, watched over by the bulging eyes of the carved heads in the wall, an assemblage of cogs and levers stands on proud display. The shiny brass plaque reads: 'William Barnestone Esqyer Gave this Clock to the parishe of Bishopstone In the yeare of our Lord 1654.' This rudimentary mechanism would have provided clarity on cloudy days. And when it malfunctioned, it would have been corrected by the shadow moving over a midday mark or a sundial – potentially the sundial up on the tower.

Even the hours on a perfectly functioning clock would have been tied to the Sun in the local sky until comparatively recently. To put a long and complicated history very simply, after William Barnestone made his gift to the parish, it would be another two centuries or so before towns and villages in the British Isles stopped setting their clocks by the local Sun.[32] That is, before we lost the everyday art of telling time by events in the sky above us.[33]

With a clank, Rosie and I close the door to the church and catch the drift of woodsmoke in the air. We stand back to look at the barely discernible lines in the grey walls, and feel strangely thrilled to have found such a wealth of dials that would have cut the day into different shapes by the Sun.

The dials are hiding in plain sight, Rosie reflects. It's far easier to notice the snarling gargoyles on the tower or the vivid zigzags around the Norman door than these much slighter scratches beside them. But if we let our eyes linger long enough to perceive their presence, we might find other meaningful marks in the stone.

There was a moment earlier this morning when the hazy light sharpened enough to cast a definite shadow on one of the dials. I pointed to where the rod would have been, taking care not to touch the soft stone. The shadow of my finger hit one of the hour lines, and I imagined another hand beside mine scoring that line many centuries ago. The sensation of our bodies so close in space yet so remote in time brought to mind another old sundial motto: *umbra sumus* – we are shadow.

The fragile etchings on church walls are the remnants of the touch of people who are otherwise unknown to us, as Matthew Champion has evoked so movingly in his book *Medieval Graffiti*. But their marks are slipping away. After the last half-century of acid rain, the weatherworn lines scratched into exposed walls have become discernibly fainter.

Like apparitions hiding in the stones, there are scratch dials mentioned in the annals that have failed to reveal themselves to me. Other sleuths, I'm sure, will have more luck. And happily, this is a hunt we can all join. A great many medieval dials have yet to be properly recorded around the UK and Continental Europe. Ben and his colleagues encourage us to seek them out and share what we find with the British Sundial Society and its European counterparts.

'Here is The City', says the sign to a winsome huddle of thatched cottages at the edge of the village. This tiny district of Bishopstone is more of a sixteenth-century hermit's idea of a metropolis than the one we left early this morning. The City has no road, just a wiggling footpath through lovely gardens. And to Rosie's particular delight, this is our route up onto the Downs.

Above the village, the hedged footpath brings us out into rolling fields shining silvery-green from the chalk soil beneath the lines of ripening wheat. Further on and up, the footpath meets the Ridgeway, the ancient track that runs along the high chalk escarpment (scarp) at the edge of the Downs, and we turn left along it to walk back to Woolstone a few miles to the east.

High up on the Ridgeway, the grassy Downs sweep around us in velvety waves under thin shoals of cloud. Even in gorgeous June, this wide-open space feels rather remote and exposed. Settlements are now sparse and humans few. Time, too, stretches into an awesome expanse. A weather-mottled board by the track says, 'People were already settled here' about six millennia ago. Its map of the land around us is dotted with ruined forts and long barrows. And as we carry on along the Ridgeway, we pass Wayland's Smithy, a huge boat-like chamber tomb built more than fifty centuries ago – long before the legend of the magical blacksmith Wayland was brought to this place.

When people believed ruins were the dwelling places of ghouls and goblins, would travellers approach these strange formations of earth and stone with trepidation? Would they inspire terror as darkness draws in? Even if you were unafraid of supernatural forces, this white chalk track guiding you through the wilderness must have been a reassuring sight on dark days and moonlit nights for many generations of itinerant labourers, traders and pilgrims.

For the weary or the wary, how to guess when it's time to turn off the Ridgeway and seek shelter in the villages nestled among the sycamores and oaks down in the vale? How to judge the hour? How to know what is left of the day before the light fails? Maybe you'd hear the church bells floating up from the villages. If each place were to regulate the bells by its own scratch dial, there'd be a wonderful lack of synchrony to the chimes. Yet the bells would be inaudible when the wind roared or blew their sound away.

If you're a well-equipped traveller, you might carry a pocket sundial. Long after this trip, I would have the lucky opportunity to meet up with Mario in Canterbury. In the misty light of a winter afternoon, we strolled through the colonnade of the cathedral and up into the water tower, where he brought me to see the Canterbury Pendant, pictured below. This silver and gold tablet was found by workers excavating the Great Cloister in 1938. It's a portable sundial pierced with holes marking mid-morning, midday and mid-afternoon in each month – perhaps for reckoning hours of prayer during the day. The brighter months of the year are inscribed on one side and the darker on the other. Mario showed me how the gnomon has a detachable peg tipped at one end with the tiny head of a gem-eyed beast holding an orb (the Sun?) in its mouth. The pendant is a simple tool, but one with awe-inspiring beauty and perhaps great symbolic and practical power to command events. By tradition, if not in fact, it belonged to St Dunstan, the tenth-century Archbishop of Canterbury. It would have gleamed wonderfully in the light as its owner dangled it from its chain to divine the time.

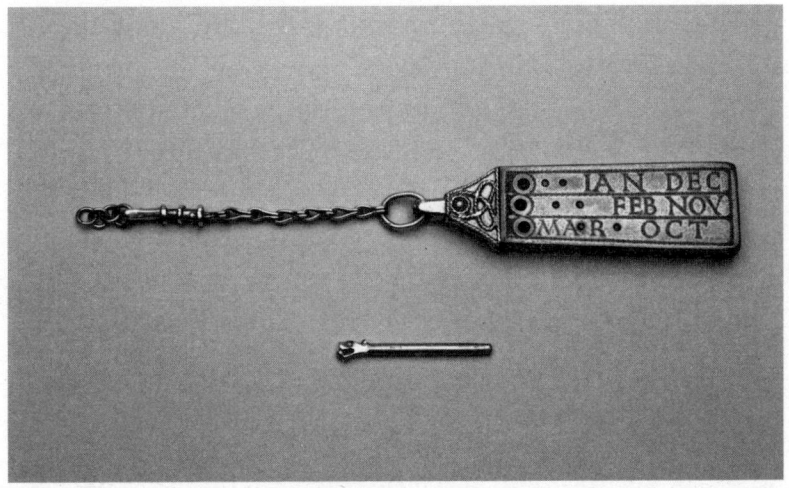

This exquisite instrument would have been an elite possession and has an unusual design. But portable sundials with other forms, and crafted from cheap materials like bone or wood, were carried by travellers for hundreds of years across Europe and the Islamic world.[34] One of the great discoveries Mario has made is to identify that a small bone cylinder found in a Roman tomb is a sundial. It belonged to a physician who would have used this rather more mathematically precise tool, he said, to know the best time of day to administer particular medicines.[35]

Here on the Ridgeway, if Rosie and I lacked a device to know the hour, we wouldn't feel completely rudderless in a sea of nameless time. Under the brightening sky, we're in the presence of many ready-made sundials. We know it's nearing time for lunch when the shadows of trees and fence posts point north, which they do every day at midday (here in the northern hemisphere). And we know they'll keep turning eastward through the afternoon.

Or, if we were unsure which way is north, another way to keep an eye on the time of day is by the *length* of shadows rather than their direction. At noon, when the Sun is highest, shadows are smallest. And at the end of day, in the charming words of Charles Cotton, shadows make 'Mole-hills seem Mountains, and the Ant / Appears a monstrous Elephant'.[36]

Middling shadows at other hours, however, are harder to gauge. That's when we might turn to the shadow-maker and measuring tool we always have with us: our body and our feet.

It's intuitive to note the stretching and shrinking of a shadow, Mario said, and it is one of the earliest ways of reckoning time. There are manifold methods across global traditions of tracking time by the shadow of your own body or an upright post. In medieval monasteries, for example, the fluctuating shadow lengths for the seasonal hours through the year were beautifully inked into semicircular tables labelled 'horologium viatorum', traveller's time-teller.[37] The unit of measure was the foot, and the body the gnomon. But the 'foot' wasn't a fixed standard as it is for us. Instead it was assumed the traveller's own foot would divide their height by a common ratio, for instance, 6:1. To know the time, the traveller would stand with their back to the Sun and measure the length of their shadow with their feet (presumably by lining up the shadow with a mark on the ground and pacing towards it, heel to toe). Then they'd compare the number of feet to the table of hours for that month. In Mario's view, the tables would be easy enough to memorise without needing to consult a crumpled sheet of numbers in your pocket.

Later, tables were devised for judging the equal hours of the clock (rather than the old seasonal hours) by changing shadow lengths through the year. Chaucer – whose characters mark time in all sorts of ways – implicitly refers to one of these schemes at the beginning of the *Parson's Prologue* in *The Canterbury Tales* (1387–1400). The narrator guesses the time is four o'clock partly by the length of his shadow measured in feet: a unit he defines as one-sixth of his height.[38]

Rosie and I live near (what is now called) the Old Kent Road, the ancient highway that presumably Chaucer's protagonists and definitely actual medieval pilgrims took out of London on their way to Canterbury. Whenever I'm crossing it, I imagine them on horseback winding through the twenty-first-century traffic and glancing at their shadows to guess the hour. Assuming they would do this, and even if their methods were rough and flawed, marking the time by their own shadow might at least feel more consistent than relying on wobbly lines on scratch dials at country churches.

Moving in the opposite direction to the pilgrims would be Kentish drovers slowly nudging their herds around the Vauxhall Corsas and Lime bikes. Out in the countryside, these herders must know when to set off for London markets, break for dinner, and bring their flock to shelter before darkness falls. And they might have their eye on shadows too. Certainly, there are reports from not so long ago that herders in Europe had their own ways of improvising sundials on the move.

Lithuanian shepherds in recent centuries, for example, would bring in their flocks at midday in midsummer when their shadow was short enough to step or leap over – which conjures a joyous image of young shepherds hopping under the high Sun. One man born in 1936 recalled how the measure of shadow length was adjusted for the season: 'When you turn out the livestock in May the shadow is five feet long, in June-July – four feet, in August – five feet...'[39] I have also come across a brief recollection from one Icelander in 1970 that people would measure their own shadow to know the time.[40]

There are other ways of using your body to improvise a sundial. Mario told me about techniques that Italian shepherds in past centuries used for mapping shadows with their hands.[41] One especially rough-and-ready method has been documented in a photo without explanation,[42] but it seems easy enough to figure out. This is Mario's sketch of it, below.

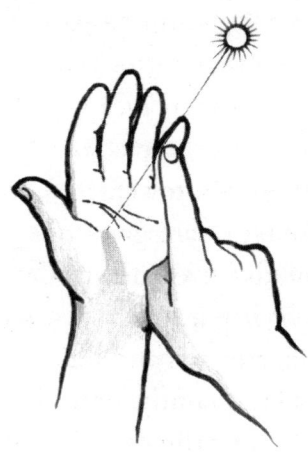

It would work if you held the palm of your left hand facing the southernmost point on the horizon (where the Sun is highest at midday), and your right index finger aligned with

SCRATCH DIALS AND STICK DIALS

it too. When the shadow on your palm is directly behind the finger, the time is around midday. When the shadow is closer to your thumb, it's still morning. When the shadow is closer to your little finger, the day is declining.

The gesture of angling your hand to estimate the hour like this feels weirdly close, I think, to checking a wristwatch. Although here, the time isn't held inside an instrument but measured by how the light falls against your hand.

Around the Ridgeway today, small flocks of dozing sheep look like tufts wafted down from the woolly sky. The fields are hedged now. But in past centuries there may have been shepherds out here moving their flocks, sometimes for long distances, to find fresh pasture. And it's not beyond reason that some of them constructed sundials in the turf.

The twentieth-century poet and historian Geoffrey Grigson asserted that sundials were 'at one time cut in the close grass of sheep-walks and hills in many parts of England'.[43] He is not alone in expecting that old field names like Dial Hill and Dial Pasture may reveal their location. (On the other hand, those names might mark the place where you could see the clock on a stable or church tower.)[44]

Unexpectedly, Shakespeare gives one of the clearest clues that English shepherds were known to construct sundials in the turf. He must have expected his audience to grasp what Henry VI means when he sits on a hill wishing he were a shepherd carving 'out dials quaintly, point by point, / Thereby to see the minutes how they run'.[45]

In 1909 the folklorist Edward Lovett reported how shepherds told time on the South Downs, another high grass-and-chalk landscape about a hundred miles away on the south coast.[46] The shepherds spent their days roaming the sheep-walks and needed to leave enough time to gather and coax their flocks to the safety of the fold before nightfall. Yet they didn't always have use of a watch or hear the chimes from a clock. So some shepherds built various forms of sundial by cutting marks in the turf under the shadow of a stick – or from an assemblage of sticks.[47]

Lovett gives a method for constructing one type of dial made by the South Downs shepherds, which is broadly this. Mark a circle about 45 cm or 18 inches in diameter in the turf. Place a small stick upright in the centre. Place a tall stick (about 30 cm or 12 inches tall) due south of the little stick. Place another tall stick due west, and five more evenly spaced in between. Here's a picture of my attempt one lunchtime a few weeks before today's Ridgeway trip.

Many simple sundials, like the ones we found earlier on church walls, have a single gnomon at the centre whose shadow points outward at an arc of time markers. The South Downs shepherds did construct that kind of dial in the turf too. But in this design, there is more than one significant shadow, and

SCRATCH DIALS AND STICK DIALS

they all point *inward* at the little stick. In the picture above, the shadow of the first tall stick is pointing north at the little stick, which means the time is around (local apparent) noon. As the Sun moves west through the afternoon, one by one each tall stick's shadow would point at the little stick and mark a different time.

I showed this photo to Mario, and he confirmed that even if all the sticks were perfectly placed to match the clock hours, after a few days the dial would disagree with the clock because the Sun's path in the sky changes a little each day. But, he reflected, like the dials made using your hand, this temporary tool may have been fit for purpose for the shepherds.

When I first followed Lovett's instructions to make the dial with sticks, I stretched out on the soft grass to contemplate its form. As an adult, how often had I paused long enough to watch a shadow move by the width of a pebble or even a blade of grass? The change was both achingly subtle and unnervingly fast.

As with many methods I'm trying out in this book, of course I have no practical need to mark time by shadows. The digital clock on the phone in my pocket is synched to ultra-precise standard time. Its old-fashioned design still looks a bit like the early kind of clock dial whose hour hand loosely maps the apparent journey of the Sun.[48] Yet the time it gives is no longer tied to my local sky. The hour hand of the modern clock, you could say, is a solidified sunbeam severed from its source.

Today it's so easy to forget how past generations understood the Sun to be the pre-eminent giver of time on Earth.[49] While my stick dial was a very blunt tool indeed compared to my phone, the process of constructing it helped to expand my sense of the measure of time beyond machine and screen.

I'll never know what it meant a thousand or a hundred years ago to mark shadows with lines in the stone or sticks in the turf. Nonetheless, the beat of time becomes something else when I study the flow of shadows and feel the warmth of sunbeams linking me umbilically to the sky. That's why it's worth lying in the grass watching shadows for a while. Their shifting pattern declares each moment to be unique, not just a unit on repeat.

When Mario and I were talking about sundials made by the South Downs shepherds, he recalled how his own grandfather, a farmer working in the Italian countryside, would come home in the evening when the Sun touched the treetops. That time of day would change through the year but, Mario said, the method worked for him.

Mario's memory of his grandfather reminded me of a story shared by my father, who was born in 1933. We were chatting about what the working day was like when he was young, and I happened to mention I assumed he'd always worn a watch. But it turned out that wasn't so. When he was young you couldn't easily get hold of cheap wristwatches that would withstand the damp, muck and knocks of farm work, my father recalled. 'Anyway, sometimes farmers don't want wristwatches because they get in the way!'

Dad remembered one man on the farm who habitually carried the time. If you were out working with him, he might fish a tobacco tin from his pocket, prise open the watertight lid, unwrap the lint cloth, take a squint at his watch, and ravel it up again. What prompted this effort? 'When things got boring,' Dad guessed. 'Or when it was getting towards his lunch.'

SCRATCH DIALS AND STICK DIALS

Most often, Dad recalled, when they were out in the fields – shearing, threshing, fixing gates, pulling thistles from the corn – they relied on chimes from the church clock. But there were areas of the farm where you couldn't hear the bells. In those places, they calculated break times by the trains rumbling back and forth between the towns.

'When the quarter past ten went by – any time between ten and eleven – we'd bring out the can of tea and have our mid-morning snack.'

'The train gave us an indication of the time,' Dad explained. And it was a pleasant part of the rhythm of life. 'When you don't have a watch on, you get your body tuned in,' Mum agreed. 'You listen for the bells or for the train bringing the children home from school.'

In the fields of the industrial Midlands, where clocks had been part of everyday life for centuries, you might still heed the time from other clues in the world. And I expect the clanking of the morning milk lorry, the cawing of the rooks and the dimming of the light would be equally obvious signs of the phase of day.

Up on the more isolated South Downs, some of the old shepherds, Lovett wrote, could 'make very good estimates of the time without either watch, sundial, or visible Sun'.[50] He doesn't say how. But on this sporadically cloudy day on the Berkshire Downs, thanks to the adventures of previous chapters, I can feel myself becoming a little more alert to changes happening around me. The most consistent sign as we walk has come from the pink-and-white trumpets of the bindweed trailing along the Ridgeway. I can't make 'very good estimates of the time' by them, but the flowers are visibly closing up as the afternoon goes on.

THE FULLNESS OF TIME

If the sky gods were to glance down right now through a chink in the cloud, they'd see Rosie and me walking off the Ridgeway toward a huge animal drawn in the chalk. This landmark, known as the White Horse of Uffington, may have first been cut into the hill more than three millennia ago. Astonishingly, archaeologists have found that its shape has not substantially changed in the care of perhaps a hundred generations.[51]

We arrive at the horse just as the newest generation, a group of children from a local primary school, has finished helping the National Trust ranger to weed and scour its surface, ready to be packed with fresh chalk. They trundle and skip back to the minibus, with the teacher's firm voice carrying above the chatter. Soon human sounds fall away, and we're left alone on the hill with the lark singing above and the red kite circling below.

Rosie and I step gingerly around the horse, surprised by its slim lines. Seen from a couple of miles away down in the vale, this leaping figure shines out from the hillside. Up close, the animal's eye is no wider than the length of my arm.

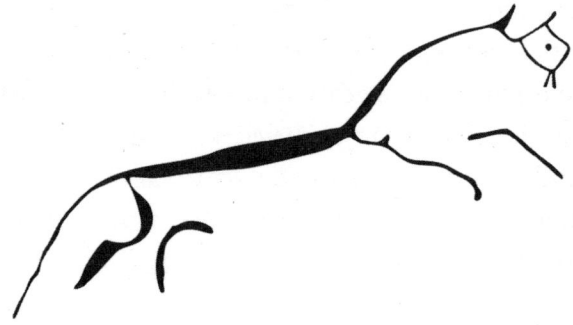

We walk back up the slope and look out over the Vale of the White Horse. In the distance are Swindon town and

SCRATCH DIALS AND STICK DIALS

a line of huge white wind turbines. Below us to the left, there's the little church at Woolstone peeking from the trees. Straight ahead, more or less, there's the sturdy belltower at Uffington, and away to our right, another village obscured by trees. All these structures punctuating the plain puts me in mind of something Lovett slips quietly into a footnote. The shepherds, he writes, also reckoned 'the time of day by the position of the sun in respect to the tall cathedral spire' in Chichester.[52] The shepherds reckoned the hour, in other words, by the Sun moving past that giant time marker at the foot of the Downs. (Hopefully, they didn't risk harming their eyes by looking directly at the Sun.)

It occurs to me now that it would be helpful if the shepherds could see several spires positioned evenly around their horizon, so they could divide the day more finely. Wouldn't it be useful, too, if they could somehow carry the spires with them as they roamed the Downs?

I wonder if that way of thinking might have influenced the design of the dial with so many sticks. It looks to me like a portable model of an ideal landscape. The short stick in the centre of this miniature world is the shepherd. As the Sun moves around the horizon, the shadow of each spire-/tree-/hilltop-stick points to the shepherd-stick.

I have no idea if this comes close to how the South Downs shepherds thought about their dial. But why wouldn't herders – who spend their days studying the pasture, the coming weather, the fading light, the best route home – make a sundial like a landscape?

Looking out from the high Ridgeway over the Vale of the White Horse, I imagine shepherds here would have had many equivalents to Chichester Cathedral spire to track the Sun as they grazed their flocks on the grassy slopes of the scarp. Then I realise that, unlike Lovett's south-facing shepherds, we're looking north with our backs to the light. All the landmarks spread out before us – the villages and towns, the woods and towers – are useless as hour markers because, from this angle, we'll never see the Sun pass over them.

On the other hand, someone down in the vale looking back at us would be aware of the Sun in the south moving along the scarp, the high northern ridge of the Berkshire Downs that runs for miles roughly east to west. And on dark winter mornings, from certain places, they could observe its dawn glow rising above the scarp's most striking landmark of all, the White Horse. There is an enticing theory that this prehistoric monument might have been deliberately aligned with the morning Sun when viewed from the landscape around Woolstone, as we'll discover in the next chapter.

5

Daymarks

Tracing time by the light over the landscape

All around Iceland there are places named for the time of day, like Miðmorgunsvarða (Early-morning Cairn), Dagmáladalur (Morning Valley), Miðdegishæð (Midday Hill), Nónfjall (Mid-afternoon Mountain), Miðaftansrúst (Evening Ruin), Náttmálafoss (Night-time Waterfall).[1] These special kinds of landmark are known as daymarks and were once common across Norse cultures. But they survived for much longer in Iceland and have left exceptionally vivid traces. When you start searching the map for names for daymarks, they readily appear.[2] In the land around the airport. On the fringes of the city. Among the icy skerries in the whale-grey waters. Under the heaps of soft cloud on the lumpen fells. Near the red lava trench sending up huge plumes of smoke. On the snowy peaks rising like silver-black ziggurats from plinths of scree.

In past centuries, each farmstead in Iceland had its own daymarks to divide the day by the Sun moving around the local horizon. If we were looking out from a particular place, when the bright glow behind the cloud arrived above, say, a mountain due south, we would know it was more or less midday. I say *we*, but of course today we would never dream of looking directly at or near the Sun, even through sunglasses, because of the risk of permanently damaging our eyesight.[3] (It's easy enough to have a sense of where the Sun is without looking at it, and we can check shadows if we want to know its precise location above the horizon.)

This trip is about six months after Rosie and I visited the Vale of the White Horse. We've leapt at the lovely offer to stay with a friend for a week in Reykjavík. I've long been fascinated by the daymarks here and at last we have the chance to come and explore them. The opportunity falls in mid-January, but never mind, I think, there are bound to be sunny moments. As it turns out, this week's weather is a surprise even to the locals. A thick, damp blanket of cloud seems permanently draped over the snowy fells. The weak blue daylight dims to dusk soon after lunch. All the morning hills and midday peaks scattered about the land feel less like a means of telling time than hopeful memorials to the long-lost Sun.

These aren't the *ideal* conditions to go looking for daymarks, Gísli Sigurðsson confirmed with magnificent understatement. Yet here we are together, stepping gingerly down a path in the snow between walls of jagged black rock.

Gísli is an eminent expert in Icelandic literature and folklore, and has, wonderfully, agreed to talk about the old ways of telling time by the light moving over the peaks and valleys. Despite the gloomy light, when I suggested we head out into the hills in a break in the weather, to my delight, he was up for the adventure.

Rosie has navigated our four-wheel hire car behind the tour buses for fifty kilometres along the treacherous road from Reykjavík. We have pulled on our boots and subarctic gear and walked out into the vast expanse of hill and moor.

The icy wind is stinging our cheeks pink and I'm losing sensation in my face as we descend through the snow into the Almannagjá ravine. But Gísli buoys us along with infectious energy. He is leading us into a very special place. This deep, rocky ravine, and the valley below, were once Iceland's great theatre of power, the site of the medieval parliament. While it is difficult to imagine right now, fundamental ceremonies here were scheduled by the Sun moving over the land.

After the last ice age, this subarctic island in the North Atlantic was one of the final landmasses of this size on Earth to be settled by people. The first Norse-led population (from Norway and the British Isles) arrived *en masse* in the ninth century.[4] Remarkably, their early society was not a kingdom ruled by monarchs, but a commonwealth made up of a few thousand farming communities governed by chieftains. Their parliament, the Alþingi, was founded in 930. At midsummer, for more than three centuries, the chieftains gathered from around Iceland for their annual assembly here in Þingvellir (Thingvellir), the Assembly Fields.

From a bird's-eye view, the Almannagjá ravine we're standing in draws a thin black line along the western edge of the shallow valley. Both the ravine and valley are on the vast Mid-Atlantic Ridge between the Earth's Eurasian and North American tectonic plates, which are very slowly pulling apart. The violence of that movement created this deep ravine – and it's still causing the land we're on to rupture. Indeed, on this volcanic island, Gísli warns, the ground is regularly torn open by rifts of all sizes, and you need to be careful where you walk.

Gísli leads us as we march slowly onward to where the narrow ravine creates a sanctuary from the worst of the wind. We stand and look over the eastern side of the ravine to the Þingvellir valley beneath and the plain beyond: a white-blue

THE FULLNESS OF TIME

land under a grey-white sky. Below us, the slender Öxará river winds through the valley floor from the mountains in the north to the lake in the south. Gísli points just below us to the shapes under the snow of the stone booths where the chieftains put up their tents. Close by is a natural enclosure where they corralled their horses. I try to picture this snowy dell at midsummer a thousand years ago. Smoke from the feast fires is softening the evening air. Horses stand cropping the grass and flicking their tails. And maybe more steeds and riders are racing over the plain towards us in a desperate effort to make it here before sunset.

The journey over mountain, moor and marsh to Þingvellir might take many days. So the chieftains needed to plan their time carefully, Gísli says, because they had to get here by a definite time or they would lose their position in parliament. According to the *Grágás* ('Grey Goose') Laws of the Icelandic Commonwealth, that deadline came when the Sun disappeared from the valley on a certain date after midsummer.[5]

The chieftains performed here as the lawmakers and lawyers. At their court hearings, inheritance disputes, say, or blood feuds were settled, ideally without violence. After the chieftains' rule ended in the thirteenth century (when Iceland came under the authority of the Norwegian king), the Alþingi continued as a court of law. Later, in the post-Reformation era, people were tried for witchcraft and adultery – and the condemned were punished with acts of unthinkable horror. Gísli gestures over to the Drowning Pool by the river as the wind moans through the ravine.

The power wielded by the chieftains might be hinted at by the Old Norse word for chieftain, *goði* (*goðar* in plural), which is related to 'god'. Preeminent among the *goðar* was the Lawspeaker (*Lögsögumaðr*), whose roles included judging the time to begin. On the Saturday after they assembled, *Grágás* states, the court hearings should start when the Sun is on the west wall of the ravine as observed from the Lawspeaker's seat on the Law Rock (Lögberg).[6]

That strict timing doesn't seem so very different from the definite schedule followed in a court today. Except, of course, we'd expect it to be set by a standard hour displayed on a device that can be transported and replicated anywhere. But here timings were defined by the unique form of this ravine and its relationship to the Sun, as perceived by one individual in one particular spot.

Where exactly the Lawspeaker sat, though, is not clear. There is a lump in the snow beside a fluttering flag that marks the Lögberg on the eastern edge of the ravine. Yet like so much about what happened here, Gísli says, we don't know the location for sure. Would such an important official, he asks, really sit in such a weather-wracked position?

What precisely the *Grágás* rule means by when the Sun is on the west wall of the ravine has been debated too. Does that moment come when low sunbeams set the wall aglow in the morning? Or when the Sun's disc starts sinking behind the wall in the evening? Or when its rays hit the wall at a certain angle at some other point in the day?

Among those who've tried to provide an answer was the Danish philologist Kristian Kålund in the nineteenth century. He consulted the local priest, Simon Bech, who went out to observe under a clear sky on 6 July 1876. Bech reported that sunrays hit the west wall at around half past two in the morning.[7]

Kålund concluded from this and other clues in *Grágás* that during the Alþingi – held at midsummer when the darkness hardly falls – people scarcely distinguished between day and night. The court sessions would carry on through the usual times of rest, he thought, and might have tied people to the place for very long periods.

But any theory like that is highly speculative, in Gísli's view, because *Grágás* is inherently cryptic about what features of this landscape it is referring to – which rock, what part of the wall – and precisely how people were meant to read the light moving over them. We don't know exactly how this theatre of power was laid out in the landscape.

Gísli leads us over to where the ravine wall curves like a ready-made auditorium. Scholars suspect this may have been the court space. Beside it, there's a podium-like mound in the snow. When I stand there and try to imagine justifying myself to an intimidating audience of chieftains, I have the strange feeling of being watched. The space is empty of people, but the lava rock is fissured and faceted with such characterful detail that it's easy to picture faces in the ravine wall. And above us on the clifftop, looming over one side of the 'court', is a peculiar shape rather like a straight-backed figure in bulky robes and a boxy hat.

'That one almost looks like a proud official,' I say.

'Yes,' Gísli answers. 'You see these things almost immediately when you start looking.' There are so many richly evocative formations of rock and earth in this volcanic landscape.

Gísli points out the other distinctive rocks dotted along the top of the western ravine wall. Maybe, he muses, events were timed by the Sun passing over them or sunbeams striking them. But we can only guess at so much of what happened at the Alþingi.

What we do know from *Grágás* is that marking time by the Sun was bound into the laws of early Icelandic society by very specific rules that apply only to the land formation here in one particular season. In a related way, for centuries ordinary people in Iceland tracked times of day by the Sun moving around the horizon. But they did so by a *common* system that was applied all around the island to each local landscape.

In the Old Norse world, the day-and-night cycle was split not into twenty-four hours but eight equal segments, each called an *eykt*. The horizon, too, was split into eight equal divisions, each corresponding to one *eykt*. The divisions were often marked by some kind of object in the landscape known as an *eyktamark* or *dagsmark*, a daymark, identified from the doorway of each homestead. Herders out in the pastures would have had daymarks, too. Most parts of this volcanic island are spectacularly well provisioned with peaks, islets, streams and valleys. And if you lacked a ready-made daymark where you needed it, you could make one.[8]

Picture a small farmstead with turf walls at the end of a muddy track in a green expanse of rolling moor. You're standing in the doorway looking out around you. The sky shines in the puddles in the yard. On the smooth slope of the hill to your left, the meadow grasses sway and flowers bob in the breezes from the valley. High up on the ridge, there's a stack of stones called Dagmálavarða (Daytime Cairn). In front of you is a flat-topped hill named Hádegisfell (Midday Fell). To your right, where the Sun moves in the afternoon, is another grassy mound puckered with dark crags. This is the nearest of three hills – each named for *nón*, a time of mid-afternoon – running more or less in a line to your southwest.[9] Halfway between Hádegisfell and the Nón hills sits the low fell known as Miðmundafell.

The sketch, below, places those four daymarks on a diagram of the eight chief times of day in Icelandic tradition.

These eight divisions, the *eyktir* (the plural of *eykt*), are measured by the Sun moving around the full ring of the horizon. (The compass directions are a rough guide. In practice, local communities may not have identified such evenly spaced daymarks. And, as we'll see, they may not have named or marked a full ring of them.)

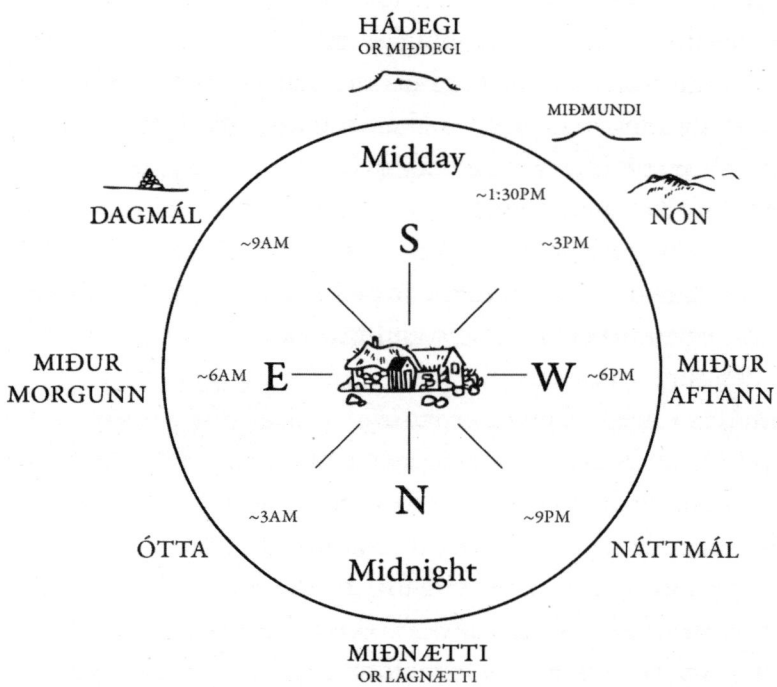

Each *eykt* lasts about three modern clock hours, and the diagram gives its loose timing. So, for example, the Sun is above the *dagmál* cairn at around 9 a.m. The midday hill is to the south. The Sun crosses over the hills named for *nón* in the southwest at about 3 p.m. As well as the major divisions, people sometimes marked smaller spans of time – and here the hill named for *miðmundi* identifies the half-*eykt* between midday and *nón*, at around 1.30 p.m.

Inevitably, hourglasses and clocks would reveal how the *eyktir* measure durations that alter through the seasons. But they don't change span by nearly so much as I'd guessed. Despite Iceland's famously extreme fluctuations in daylight, the Sun appears to turn at a more constant rate around the horizon (and closer to it) than further south. In his evocatively titled article, 'The Subarctic as a Sundial', the historian of science Thorsteinn Vilhjálmsson calculated that at the latitude of Reykjavík (64° North), time measured by the Sun reaching the same point on the horizon differs from the clock by only up to two or three quarters of an hour through the year. That mismatch may 'have been quite tolerable', he remarked, for an agricultural society.[10]

When the German folklorist Konrad Maurer travelled through Iceland in 1858, he learned from a priest that people identified the daymarks in the landscape by the location of the Sun on the longest day.[11] That would be the obvious time of year to select the daymarks, when the Sun is above the horizon for much of the day and people are most often outdoors.

On the map, above, south sits at the top because this is where the Sun peaks in the middle of the day at *hádegi*, literally 'high-day'. Likewise north sits at the bottom because that's where the Sun appears to be at midnight, *lágnætti*, 'low-night'. I had guessed this low-night point on the horizon was only theoretical because the Sun never appears there. But in many places in Iceland in high summer, Gísli elucidates, the Sun does make a visible circuit around the entire skyline. 'At midsummer,' he says, 'the north coast gets a midnight Sun that never sets – it just dances on the horizon. Even as far south as Reykjavík, where the Sun disappears for quite a while at midnight behind the mountains, you can tell where it is by the glow behind the peaks.'

Imagine standing in the yard of that homestead in the moors and noticing the nocturnal gleam gliding along very slowly behind the mountains. Everyone else is deeply asleep and you leave them to rest. But one night soon maybe you'll be out here watching for when the glow reaches a tabletop mountain close to the northeast. And then you'll wake the community, because this is the season to get up early – to make hay while the Sun shines, quite literally.

Hay is rich meadow grass cut and dried to sustain the herds through the frozen winter. There is a saying in Iceland: farming is about haymaking. I was told this later by Halla Steinólfsdóttir, who farms sheep and goats on the northwest coast. 'If the hay is good,' she said, 'then everything is good.' And to be good, the grass must be cut on time and it must be dry. That means haymakers must make the most of every moment and act fast before the weather turns. According to a commentator in the 1830s, in Iceland 4.30 a.m. was 'the rising time of the workpeople in the hay making season'.[12] Still today, haymakers get very little sleep.

There is much about farmwork that flexes with the seasons. And in eras past, Icelanders adapted the clock to flex along with it. Sigrún Kristjánsdóttir has written about traditional practices of reckoning time in Iceland. She told me that when clocks first became available, 'people set them as they pleased'. During haymaking in the summer, people would fix their clocks up to three hours early to make the most of the light.[13] The practice is much like when our standard clock changes to daylight-saving time, except in this case the clock is reset locally to serve a particular purpose.

DAYMARKS

When we're walking by the river at Þingvellir, I ask Gísli about how rural Icelanders would have made the transition to standard clock time. There are stories, he recalls, about farmers struggling with the difference between their own domestic time told by the Sun and standard time after radio broadcasts were introduced at the end of 1930. They were keenly interested in the weather forecast but complained about having to get to the radio at the right moment to hear it. The farmers, he thinks, 'were not accustomed to the idea of thinking in minutes'.

Gísli knows well the rhythms of agricultural life, having worked on a farm in his youth. The most important aspect to time-reckoning on a farm, he reflects, is to make the most of the sunshine. 'That's when you take timing very seriously, especially if you're haymaking.' Farming remains a craft of timing by the weather, the dew and the light.

We head back to Reykjavík when the feeble day starts to fade. Somehow Rosie remains calm at the wheel despite the amazingly capricious weather: in quick succession, the car windows clatter with hail, smear with sleet, clot with snow. In the frozen fields, all warm-bodied beings seemed huddled away deep under shelter. All but for the sturdy Icelandic horses pulling lumps of fodder from the big round bales slowly slumping in the snow.

When you're hunting about on the map of Iceland, it's easy to find a midday hill over here and a mid-afternoon peak over there. But locating a substantial set of daymark names clearly linked to one place is a much bigger challenge. That imaginary homestead, pictured on page 120, is loosely based on a real site with four daymark names on the map, all concentrated around the middle of the day. But that's an unusually high number.

Why is it so hard to find a fuller set of daymarks on the map?

For a start, most likely, people didn't identify daymarks for all the *eyktir*, but only the ones they found most useful. The archaeologist Birna Lárusdóttir has analysed a large set of daymark names and found the most common are for *miðdegi* and *nón*. Both mark times in the middle of the day when humans are most busy, she points out, and when the Sun is most visible in winter.[14]

Another reason for the absence is that an object in the landscape doesn't need to be named as a daymark for people to have used it to tell time. And then there are the daymark names that did exist but haven't survived. Gísli has found a report from 1959 of a farmer, Methúsalem Methúsalemsson at Bustarfell in the northeast, who could remember a complete arc of five daymarks, running from Dagmálabotn in the morning to Miðaftansbunga in the evening.[15] His family had been there for centuries, Gísli said, and their history was still ingrained in the landscape. But when farms change hands, especially in the age of the modern clock, the old names could easily slip away for those three trickling gills on the fell, say, or that birch spinney by the track.

One more likely reason for gaps on the map is that the same hill or stream could have been used to tell time by neighbouring communities with different viewpoints. For example, one small farm's Hádegisfell might be another's Náttmálafell. It would be no surprise if the full variety of hyperlocal names for a particular landform are not captured on the chart.

The names that are written plainly on the map hint at what else might have been. And not only in Iceland, because there are signs of 'sun marks' elsewhere in Europe and beyond.

For example, there is Nonsteinen (Mid-afternoon Stone) in Norway, and the area of Undersåker in Sweden, whose name may have come from a field lit by the Sun at the daily time of 'undorn'.[16] From a hunt around the UK map, Noon Sun Hill near Manchester seems especially likely: it sits just south of a village on the western side of the Pennines. And there is speculation that a boulder called Noon Stone, on a hill near Ilkley in West Yorkshire, got its name from its role as a sun-time marker.[17] In Central Europe the most famous mountains named for times of day may be the neighbouring peaks in the Dolomites known as the Sesto Sundial. These are named by number from Peak Nine (Cima Nove) to Peak One (Cima Una) and would have been used to reckon time in the village below.

In 1980 the historian Nina Gockerell lamented how 'modern cartography has frozen once and for all' those names 'that were valid for only one point in the valley, so that today numerous Midday Peaks' (and the like) 'bear names that are meaningless, for one side of a valley at least'.[18] How splendid it would be to have a map that captures all the old daymark names – that traces the flow of light over the land from every point of view.

The age would come when people no longer rebuilt the daymark cairns or gave noon peaks a thought. But that may not have happened as soon as the clock arrived. Instead in some places, marking time by the Sun moving around the horizon appears to have continued in tandem with the clock, at least at first.[19] Among older Icelandic people surveyed in 1970, one man remarked that the method of telling time by the Sun over the daymarks had been abandoned because it no longer matched the clock which had changed to 'English time'.[20] Presumably he meant in 1908, when Iceland's clocks were officially set to one hour behind Greenwich Mean Time.[21]

Even later some Icelanders may have held onto the old method, especially when working outdoors without a watch (like my father in his youth in the last chapter). The ethnologist Árni Björnsson is among the generous Icelanders who have helped me understand the daymarks system. He told me that when he was born in 1932, most households had clocks. But he did have a 'vague recollection as a child of old men working outside and looking up at some point in the landscape, saying something like: "Well, it seems to be time for coffee."'

The idea of measuring time by the glow above a mountain peak or the shadow of a cairn seems very far removed from life lived in London's tall streets. When I'm back home from Iceland, I lapse into thinking that way of sensing time belongs only to rural places in the far north in the remote past. But then I remember the video call I had with Birna the archaeologist. Can't you see your own daymarks, she asked encouragingly, in the city around you?

Yes, indeed, I realise. Years ago, I'd mapped the time when the Sun appears at its highest point in its daily arc ('local apparent noon') above different locations across London. For example, midday by the Sun happens at Hammersmith Bridge in the west about one minute after the Royal Greenwich Observatory in the east. And because shadows at midday always point due north, I included on my map highways with long sections running north to south. By chance, I live close to one of those stretches of main street and, after making that map, became almost euphorically sensitive to when the light floods straight up the road. Now, thanks to Birna, I realise this is my midday-mark, my Noon Street.

But can I call sunbeams streaming up a road a daymark?

In its broadest sense, *dagsmark*, daymark, might mean anything that helps you judge the time of day.[22] In that case, my office computer is an unintended daymark, because its display gets bleached by the high noon Sun. You may have mapped your own daymarks if, say, you've moved your breakfast chair to catch the first rays of morning or hung your washing in the spot that's warmed by the beams of mid-afternoon.

In London, the Sun's apparent journey around the horizon is not as constant through the seasons as it is in Iceland. Except that on every (sunny) day of the year, as I've mentioned, the noon shadow will always flow north straight up Noon Street. That makes this my most steady daymark and, as a novice, the one I notice in everyday life.

If you don't need to divide time at all finely, a midday-mark would be a useful rough guide to the passage of time through the whole day. It was once a custom to mark the noon shadow of a window frame or post on a south-facing sill or porch. In his book on sundials published in 1973, Albert Waugh wrote that the practice was common among the early European settlers in North America and that the marks were still to be found in old houses.[23]

The midday shadow could be measured to keep a grasp, more or less, on the calendar, too. In 1759, *The Gentleman's Magazine* of London reported that at the winter solstice the countryman marks where 'the shadow of the upper lintel of his door falls at 12 o'clock'. Then on New Year's Day, 'he finds the shadow is come nearer by the door by four or five inches, which ... he calls a cock's stride'.[24] (That's a fine old expression for how much the midday shadow shrinks between midwinter and New Year.)

Waugh mentions an alluring tradition of making a midday-mark on the floor with metal tacks,[25] and I imagine how they would polish to a gleam underfoot in the doorway of our home. Sadly, though, midday rays do not reach into our north-facing flat. But down on Noon Street, in the weeks when the trees are starting to leaf up, I note with pleasure how noon shadows are beginning to shorten. Later in the year, when the leaves have darkened and rustle in the breeze, noon shadows are palpably lengthening again. Another sign of the changing seasons is how the location of sunset moves back and forth along the line of buildings on the west side of Noon Street.

The places where midsummer or midwinter sunrise and sunset happen on the horizon depend on your latitude. At British latitudes, the rule of thumb is that the Sun's rising and setting points are northeast and northwest at midsummer and southeast and southwest at midwinter. From my favourite spot on the street, around midsummer the Sun descends behind high towers at the north end of the road. On midwinter afternoons it flames behind the gym to the south. This is an imprecise measure yet a much more sensational, theatrical sign of time and season, I think, than the subtle shrinking of a midday shadow.

Marking the location of solstice sunrise or sunset on the horizon is something humans have been doing for an astoundingly long time. In Western Europe the best-known prehistoric structures oriented to solstice are Stonehenge in England and the Brú na Bóinne complex in Ireland, both built around five thousand years ago. Famously, the central axis of Stonehenge runs between midwinter sunset and midsummer sunrise, as pictured below.[26] At Brú na Bóinne, the Newgrange passage tomb (burial mound) is positioned so the rising Sun beams through its 'roofbox' and illuminates a long passage and inner chamber at midwinter.

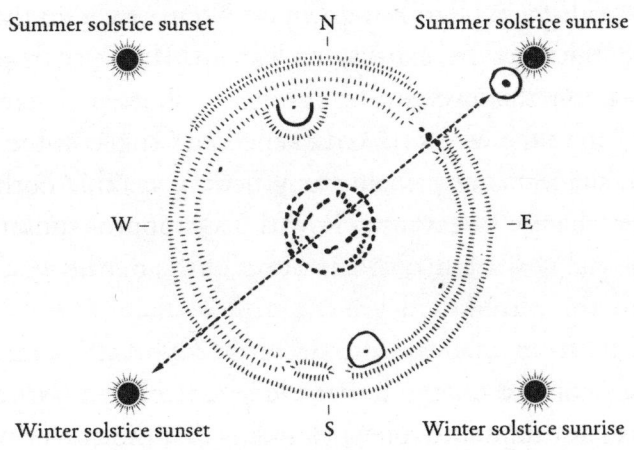

Marking key points in the solar year would help people anticipate the seasons of ripe berries and bountiful deer and the dark days when the land and water are bare and frozen. But if you simply wanted to keep track of the calendar, the archaeologist Josh Pollard explained to me, you wouldn't need to craft such elaborate, magnificent structures. These prehistoric structures remain profoundly mysterious, but scholars believe they must have been hugely important ceremonial sites. Perhaps their makers, Josh said, were deeply concerned about ensuring the cycle of the Sun carries on after midwinter and the world doesn't descend into darkness. The monuments and the rites performed there could have been designed as ways of actively working with or influencing the world to keep everything in harmony and good order.

We don't know what their makers believed about celestial events or whether they shared anything like our concepts of time. We can only seek clues in the stone, wood and earth, and their orientation to the land and sky. Take, for instance, what happens inside Newgrange, Josh said. The tomb doesn't just identify the rising Sun at the darkest point of the year. It draws the light *into itself* – as if the tomb has been designed to absorb (what its makers may have believed to be) the divine power of the sacred Sun.

I got in touch with Josh because of the remarkable hypothesis he has developed about the White Horse that Rosie and I visited in the last chapter.[27] This huge figure was cut into the steep slope on the northern scarp of the Berkshire Downs around three thousand years ago. And there is something about its form that had puzzled Josh for many years. One long-established theory is that the White Horse 'badges' the territory for a particular social group. It can indeed be seen for miles. Yet, if it was meant to be a kind of banner, there are better sites for it on the scarp. Instead, the Horse is positioned at such a strange angle that it's best appreciated from the sky.

Josh started to notice curious details about the White Horse and the land around it, and to piece them together. He observed how the upward orientation and leaping form of the chalk figure implies movement. And that it's placed where the ridge begins to rise in a shallow arc. And how, from certain points around Woolstone at midwinter, the Sun appears to emerge just to the east of the horse and stay close to the ridge all day. The effect is especially striking, he found, from the vantage of Dragon Hill, the oddly flattened mound in the fold of the scarp that sits like an enormous podium just below the horse.

Josh's theory is that the chalk figure honours the daily journey of the Sun — and might even be intended to encourage it. The horse could have been created, he proposes, as an effigy, an active image, of a sky horse that runs up with the solar disc as it rises from its lowest point at midwinter.

Around four millennia ago, images and beliefs about sky horses carrying the divine Sun are thought to have travelled to (what we now call) northwestern Europe from India and Iran with the domestication of horses. Actual horses may not have grazed these slopes so long ago. But the idea and image of the horse could have been carried here in tattoos on the human body and in motifs on jewellery or clothes — and with the stories that were told with them. Each community would adopt that iconography in its own way — in this case by carving one of the oldest and largest figures to have been cut by hand into the surface of the Earth.

The people who made this figure, Josh argues, might have understood the Sun to travel west over the ridge and drop down to the underworld at the dying of the day. That would match motifs of the solar cycle found on objects from other parts of Bronze Age Europe and beyond.[28] In this cycle, the Sun is carried by horses into the sky in the morning and down again in the evening, then ferried by boat through the underworld at night. Josh proposes the creators of the White Horse may have pictured the Sun descending into the underworld through the 'portal' of a pre-existing tomb, the one now known as Wayland's Smithy, further west along the scarp.

In the winter after our Iceland trip, Rosie and I went back to Woolstone and stood on Dragon Hill just before sunrise. As the dawn glow brightened, the racing clouds behind the White Horse looked like smoke and fire. And when the first sunbeams burst from the ridge, the violent dazzle triggered us to turn our backs. Facing north, we admired our crisp new shadows pointing over the flat top of Dragon Hill to the Vale of the White Horse below. The golden light over the distant fields was slowly moving towards us and shrinking the shadow of the high scarp, tangibly turning the world before us from night into day. Next morning the Sun's arc would be a little higher and wider, and would keep on rising higher over the ridge until midsummer.

It was a strange and impressive sensation to imagine the Sun's journey through the sky to be physically marked into the landscape around us, as if linking the earthly and the celestial tightly together. And I remembered when I'd had something approaching this feeling two summers before.

I did not expect to experience the sublime in Milton Keynes, it is true. Stonehenge at midwinter sunrise, maybe. Walking up a street in a 1970s new town at the start of an ordinary July day, well, the prospect seemed unlikely.

Yet around 5 a.m. that Monday in the summer of 2023, I had emerged from an underpass to be met, as if face to face, by a glowing orb suspended in a film of yellow light. I felt cowed by its force and instantly lowered my gaze from the blaze with its literal power to blind. The street all around me was golden.

My morning had begun about an hour earlier when I gently closed the door of a guesthouse in a residential suburb and walked into town. The suburb had been built around an old village, and I passed thatched cottages and a white horse standing in a field silvered with dew. In the half-light, I picked my way through roundabouts and winding paths by following the glow on the northeastern horizon. I knew the light would lead me to the rectangular grid of streets at the heart of the city – Central Milton Keynes (CMK). That's because the spine running through its core, Midsummer Boulevard, points broadly at midsummer sunrise.

By the time I reached CMK, the cobalt dome of the sky had brightened to baby blue, pale pink and lilac grey with just a scrape of cloud like soapsuds on an opalescent bowl. The general effect was reflected in the mirror-glass box of Central Milton Keynes train station, as if this herald of Midsummer Boulevard had dressed in glittering dawn apparel, ready to be admired.

Further on up the long, wide boulevard – despite the lines of trees and the elegant Modernist bus pavilion – the strip felt arid and canyon-like with so much of it hard-surfaced, plant-free and dedicated to parking. On this Monday morning, the café-bars styled like a beach shack or a mechanic's workshop at the base of the office blocks looked like fun socks on serious business suits.

What occupied me most on my journey was how the pedestrian path kept drawing me down into underpasses. At first this was frustrating: I wanted to see the dawn sky. But as I pressed closer to the high point of Midsummer Boulevard, the under-and-over rhythm started to feel like a veiling and unveiling, as if designed to heighten appreciation of the light.

At the end of the road, I lost the path but trusted the trajectory. After skirting a weedy forecourt, I found the bridge (over the deep trench carrying roaring cars) into the park. The path ahead led to a grass mound, a belvedere, topped with a tall white obelisk marking the place on the horizon where the Sun rises at summer solstice.[29]

This needle, glowing gold on one side and pointing up at the sky, suggested a miniature *axis mundi*, a monument meant to connect the Earth to the celestial realm. I stood beside it and absorbed the dawn, always averting my gaze from the violent power of the awesome Sun.

The hazy fields beyond the city seemed entirely golden-green, with thin mists draped between the outlines of hedgerows, farms and villages. Standing alone at this highpoint at that moment felt like looking into a kind of paradise.

This was soon after our first Woolstone trip, when I was starting to think about the landscape as a kind of sundial, following Thorsteinn's idea.[30] That's partly what drew me to see what happens at dawn on Midsummer Boulevard. But I couldn't get there until July, when I knew the Sun had begun to rise at a more southerly point on the horizon than its midsummer extreme – and I was prepared to be underwhelmed. Yet I don't think I'd ever had such a strong sensation of the solar year tipping back from midsummer toward winter.

The town of Milton Keynes was founded in 1967 on a rural site in central Southeast England. The aim was to provide people with a decent environment to live. And there were exceptionally high ambitions for its pioneering design. What appears to have begun as a vision of green squares – and light-filled marble buildings for the people – has been muddied or abandoned by various forces over the decades. Yet CMK has remarkable buildings and extraordinary bones.

For me, the most astonishing quality of CMK is its solar orientation. After my visit, I spoke with Stuart Mosscrop, one of the town's original architects, about how this came to be. The centre of Milton Keynes was to be laid out in an orthogonal grid oriented by the boundaries given by the railway and canal. Then the team realised it would take 'just a nudge' to align it with midsummer sunrise. Wonderfully, the idea was approved.

Remember, Stuart said, it was the late 1960s and early 1970s, and there 'was a preoccupation with ley lines, ancient routes' and the like. This team of architects, most of them young, would party in the countryside at solstice. And, tellingly, the parallel boulevards running either side of Midsummer Boulevard are named for Avebury and Silbury, places with extraordinary prehistoric structures.

Surprisingly to me, the grid is a very old architectural form. While the team laying out Milton Keynes were influenced by modern American gridded cities like San Francisco, in their research they found examples from ancient China and Greece, Stuart said. The grid is a form he greatly admires. 'Space and time very beautifully come together with the right angle,' he told me. It's easy to know where you are in a grid of streets because they have an obvious compass direction. And, he said, 'You can tell roughly the time of day by the position of the Sun in relation to the street or the avenue.'

To orient the grid to midsummer sunrise, he explained, is a way to relate it 'not just to the surface of the planet', but to the celestial events happening above. To the sky, that is, under which the whole world turns.[31]

On the morning of my visit – as I retraced my steps slowly back down Midsummer Boulevard – I realised I was heading literally downhill broadly in the direction of midwinter sunset. I was walking down from the zenith to the nadir of the year.

When I arrived at the base of Midsummer Boulevard, the mirror-glass box of the train station was shining with the vivid blues of the daytime sky. The exterior of this building, Milton Keynes Central station, was co-designed by Stuart, and I now realise how significant it is. I'd often travel through or past it and think only of its function. On a fair-weather day, how much more pleasurable to imagine it as the glistening threshold to a sky-city oriented to the Sun.

The King's Mirror (*Konungs skuggsjá*) is a book of practical and moral advice composed in around 1250 in Old Norse as a dialogue between a father and son. There are stories to teach the son about the safe season to sail where the personified Winds dress up in glittering solar robes in summer and squall or sulk in a cloud-hat in winter. And there are lessons on geography and the divisions of time. Over the twenty-four hours that make up 'a night and a day', the father explains, 'the sun courses through the eight chief points of the sky'.[32] Time and direction, in other words, are linked properties of the land and firmament, and they require a sensitive eye to discern. 'Observe carefully,' the father advises, 'how the sky is lighted, the course of the heavenly bodies, the grouping of the hours, and the points of the horizon.'[33] Telling time is part of a subtle art of perceiving a web of connected signs in the world around you.

The Night-time Waterfall and the Evening Ruin and all the other daymarks scattered about Iceland are monuments to this older way of sensing time. To navigate the day, you look out to see how the light moves over the valleys and peaks. That's quite a different way of relating to your environment from checking a timepiece.

The archaeologist Birna Lárusdóttir makes a fascinating observation about how the clock orients our mind and body. 'The person who read time by the points on the horizon stood at the centre of their landscape,' she writes. They looked outward to perceive what's happening around them. But our timekeeper, the clock, 'has become the centre point, the subject'.[34] To follow one implication of Birna's insight, when we want to know the time, we turn away from the world. But of course we don't need to abandon the clock to turn outward again – to become more alive to the rhythms of the world in which we're immersed – even in the city. To stand, that is, inside our own sundial.

It was Birna who encouraged me to look for daymarks in my urban environment. That process has helped me open my sensory 'antennae' to enjoy more of the tiny events around me. Even when surrounded by digital clocks, the slow slide of sunbeams over an office floor has become intriguing. Just to notice the shift in the shadow of a building on the wall opposite is to feel, I think, the 'liveliness' of the universe. And perhaps picking up on signs of the time like this is a way to feel more connected to the rest of the planet and cosmos too – because the sunbeam that just set my Noon Street aglow is moving on to the next neighbourhood, and the next, and on and on out over the sea.

THE FULLNESS OF TIME

When we stand looking outward, naturally not all the daymarks, in the broadest sense of the word, are as bold and clear as a bright shard of light on Noon Street. Some are as subtle as a growing silence or a change in the hues of the air.

Back at Þingvellir, when we were standing in the snow under the cliff, I asked Gísli what the night is like in Iceland at midsummer. The 'magic light hangs on for hours and hours,' he said. There are no stars. The atmosphere vibrates with colour, sounds fall away, and the wind seems to calm. 'It's not just one sense that is working. You see it with your eyes. You hear it with your ears. And you experience it with the movement of the air around you.'

There's an expression for twilight, *rökkurró*, that Gísli associates especially with August when the darkness begins to return to the night. *Rökkur*, he said, means the 'half dark' and *ró*, the 'quiet'. *Rökkurró*, he repeated slowly, so I could hear the cascade of rolled Rs and hard Ks landing on a soft 'oh'. '*Rökkurró* would be the calm of this almost dark condition.'

When the almost-darkness takes hold, and the Sun drops away, how to mark time then? Once the night floods in, if you lack a clock, does the measure of time dissolve too? This is what we're on a quest to discover in the next two chapters. First we're going to explore ways of discerning the fleeting phases of twilight by the colours and qualities of the light. And then, when darkness falls, how the phases of the night are revealed by the stars.

6

The Gloaming and the Dimpse

Sensing time by the colours and qualities of twilight

It's early January 2024, just before we go to Iceland. For much of the last few weeks, southern England has been under a miserable cap of grey cloud. Now the sudden triumph of the Sun has sent me racing from London on the train to Somerset before the sky closes over again. I'm heading to meet Steve Bell, an astronomer with an exceptional understanding of the changing colours and qualities of the twilight atmosphere. Twilight has largely been a vague and an uncertain entity to me. In great contrast, since the early 1990s, Steve has been going out into the fields to study closely what happens as day turns to night. And this evening, brilliantly, he's invited me to join him to gain a grasp on the grand structure and tiny details of twilight.

In the bare winter landscape beside the railway tracks, the yellow sunbeams of late afternoon spangle the ragged seedheads and flicker hypnotically in the trees. It will be a while yet, I think, before the Sun sinks away and the twilight – the *gloaming* – draws in. That word, gloaming, sounds like a glooming and a moaning – a darkening of the spirit as well as the sky. But gloaming is likely to have come from an old Germanic root, *glô-*, which also gave us the word 'glow'. It may have referred to the gleaming of sunset (or sunrise) before coming to mean the dim and shady twilight.[1] And yet this audible 'glow' in gloaming still feels apt. Because at day's end, of course, the world doesn't just fade to black as if a celestial dimmer switch is being turned down. In the prelude to the night, as the light starts to drop, the sky burns with its most intense colours of the day. Hand in hand with the glooming, you could say, comes the glorious glow-ming. And this evening we're going to explore both aspects of twilight – the flaring colours of the sky and the darkening atmosphere on the ground – as the day dissolves into night.[2]

I'm cutting it a bit fine, but thankfully Steve is ready for me when I reach Taunton station, and we dash straight off to his twilight lookout. As we weave through the narrow grey lanes striped by yellow light, we talk about why the daylight takes on warmer tones as sunset nears.

Sunlight is made from a spectrum of colours, some of which are more easily scattered by air molecules in the atmosphere. The shorter wavelengths of the violets and blues are more likely to be scattered than the longer wavelengths of the reds and oranges. During the day, that's why the sky seems blue.[3] But when the Sun is low to our horizon, its rays travel to us at a different angle and pass through a greater depth of atmosphere. Given that longer journey, the violets and blues are scattered away much more, leaving the oranges and reds to dominate.[4]

The particular colours of light that reach us, though, depend on the state of the atmosphere. Haze from urban pollution, for example, can veil the vibrancy of sky colours. Whereas clouds may catch and intensify the fiery rays of the falling Sun.

No two sunsets are exactly alike or predictable in all their fine details. Along with the nature of the matter suspended in the atmosphere, fluctuations in temperature, wind and cloud – and the presence of moonlight and light pollution – can all alter the picture, at least a little bit. Yet despite the ever-changing details, in general the major events are more or less consistent. Evening after evening, as the Sun sets and the light dims, the flow of colours in the atmosphere and the deterioration in the visibility of objects around us follow roughly the same pattern.[5]

Evening twilight begins at sunset. It ends when the atmosphere can become no darker, and all the stars have come out (on a clear night). Theoretically this happens when the Sun drops 18° below your horizon. And there are two twilights every day. Morning (dawn) twilight starts when the rising Sun reaches 18° below the horizon and ends at sunrise. So, while the twilight that ushers in the night may appear very different from the one that repels it, by scientific definition they're essentially the same phenomenon, unfolding in reverse.

Steve has honed his expertise in twilight as a senior astronomer at HM Nautical Almanac Office. He's a key member of the team responsible for producing the yearly *Nautical Almanac*, the star manual for marine navigators, in collaboration with the US Naval Observatory. The *Nautical Almanac* has a long history with twilight, because this is a critical time of day for astro-navigators, as we'll see. Back in the 1937 edition, the *Almanac*'s tables for sunset and the end of evening twilight were expanded to include times in between those two points.[6] This simple, versatile rule of thumb splits it into three steps – civil, nautical and astronomical twilight – as the diagram shows, below.

Each division of twilight can only be a rough guide, but broadly represents a major shift in what you can see in your

SUNRISE/ SUNSET	CIVIL TWILIGHT	NAUTICAL TWILIGHT	ASTRONOMICAL TWILIGHT	NIGHT TIME
Sun's upper limb (edge) crosses the horizon	Between sunrise/sunset and Sun's centre crossing 6° below horizon	Sun's centre 6°-12° below horizon	Sun's centre 12°-18° below horizon	Sun's centre more than 18° below horizon

environment. In layperson's terms, and assuming ideal conditions (and no moonlight), when the Sun lowers to around 6° below your horizon at the end of evening civil twilight, typically you can still see large objects on the ground but no longer distinguish details. In the sky the brightest stars and planets have appeared. When the Sun reaches around 12° below, at the end of nautical twilight, the atmosphere is likely to be too dark to see the horizon, and the sky shines brighter with amassing stars. At the end of astronomical twilight (when the Sun is 18° below), the faintest stars have come out and the atmosphere is as dark as it will get for the rest of the night.

The subdivisions of twilight seem obviously to have been added to the *Nautical Almanac 1937* to help astro-navigators anticipate when to measure the stars against the horizon to find their position on the blank expanse of the open ocean.[7] The fixed stars appear to wheel around us at a constant speed because the Earth spins at a steady rate. Simply put, if you can compare the position of certain stars (or other celestial bodies) in your sky with their predicted position at the Greenwich Meridian at the same instant in time, you can convert the difference to calculate your longitude. To make that principle work in practice, however, is a very fine and complex art requiring years of training. And you need specialist kit: a sextant for viewing and measuring the angle of stars by eye, a very accurate marine clock called a chronometer, and precise star tables like those in the *Nautical Almanac*.[8]

Critically, the astro-navigator doesn't have all night to go out and measure the stars against the horizon, because they're both visible together only during the darker part of civil twilight and in nautical twilight. Depending on where you are in the world, you could more easily miss your opportunity. In the tropics, the window between sunset and the end of nautical twilight, when the horizon disappears, is typically just under an hour. Much closer to the Earth's poles in summer, the atmosphere may not darken enough to see the stars. (For example, if you're near Lerwick in the Shetland Islands, which is about 60° North, the Sun doesn't drop below 7° between early June and early July, during the season of bright nights known so evocatively as Simmer Dim.) Wherever the navigator is, the *Nautical Almanac*'s twilight tables help them predict when, in clear conditions, they should be able to see both the stars and the line that divides sea from sky.

The art of marking pivotal points in twilight by the visibility of objects is not remotely new among global traditions, and we'll reflect on some of those methods as the light dims. Moreover, the dark limit of twilight applied by the *Nautical Almanac* had been in use for many centuries. In the late fourteenth century, Chaucer explained to 'little Lewis my son' that the start of 'true' night happens when the Sun reaches 18° below the horizon, in accord with earlier sources from the Islamic world.[9]

The *Nautical Almanac 1937* appears to have first introduced the name and concept of nautical twilight, which it combined with the existing civil and astronomical twilights.[10] The accompanying narrative explained that it's helpful to call the three subdivisions by 'distinctive' and, we might add, memorable names.[11] Indeed the choice of 'nautical' is both apt and evocative, given the importance of this time of day for astro-navigators at sea. And in combination, I think, the trio of names is richly allusive. For me, *civil*, *nautical* and *astronomical* recall the ancient hierarchy of the cosmos: what Chaucer called the 'threefold world' of 'earth and sea and heaven'.[12] Read as a poetic sequence of states, *civil*, *nautical* and *astronomical* suggests twilight begins politely before sailing from safe harbour into the wilds of open sea and starry sky.

This way of marking and naming the stages of twilight is used by many people now, from weather-watchers to photographers. These days, too, the Almanac Office is a source of information about twilight times for a very broad range of groups alongside mariners. Its data services support religious communities whose sacred timings are linked to celestial events, for example, Islamic prayer times. Sports-tournament organisers ask for guidance about how long after sunset a match might feasibly run on a certain date before it's too dark to play. The police and legal profession call on the Almanac Office as an expert witness.

A typical enquiry from the police is whether the Sun could have been low enough to dazzle a driver at the time of a traffic accident. In rarer cases, they might ask Steve and the team to advise on the finer subtleties of visibility in low light. Criminal investigators may be seeking to establish, for instance, whether it's possible the witness could have recognised a suspect at the time of the crime.

I had a mental picture of Steve in a police investigation room, pointing at complicated graphs and diagrams. But he's keenly aware, he says, of the risk of confusing non-scientists, and tries to keep the explanation clear and simple. And while the Almanac Office team refers to the standard scientific works, he tells me, some of their expertise comes from personal experience. His regular routine of observing twilight over the decades, he says, has helped him build an understanding that goes beyond the page. This deeper grasp means he can more easily guide the non-expert in what happens in twilight in ideal conditions: how the sky changes colour, when the stars come out, and how much detail we may be able to see in our immediate and distant surroundings. Happily, this evening that non-expert is me.

Steve pulls up the car beside his twilight lookout: a west-facing gateway to a cattle paddock a few miles from the sea. We squidge through the mud to the grey metal gate and survey the scene. The green pastures are criss-crossed with brown hedgerows. A barn roof pokes up from behind the trees. The distant hills are warm grey, smooth and low, like whales breaking the waves. The sky is pale blue and gold, with thin wisps of white and pinkish cloud. On the ground around us, the air seems dimmer and bluer. But my sense that it's still daytime seems to be endorsed by a fluttery commotion of long-tailed tits, who briefly settle and whistle in the hedge beside us.

If Steve and I were looking out over an unobscured sea horizon, we'd be able to observe the Sun's upper 'limb' (the edge of its disc) appearing to slip away into the sea and know that twilight had officially begun. In any other place, the horizon line is obscured by land. Here in the Somerset pastures, the Sun has already fallen behind the hills. That brings us two advantages. We're not too dazzled by the glare to see the subtleties of twilight, and we're not risking damage to our eyesight. And while we can't tell precisely when sunset happens by the modern astronomical definition, presumably that wouldn't have concerned past generations of laypeople, for whom sunset was a crucial time marker. In official orders for medieval London, for example, the night watch guarding the City gates was to start at sunset.[13] That must have meant when the Sun dropped below the local skyline.

When the western sky is veiled by cloud, there are other signs of the time to watch out for. As the Sun drops below the horizon, Steve explains, the Earth's shadow begins to show in the opposite direction. (This bluish-grey 'dark segment' appears when the atmosphere closest to the Earth's surface scatters away the light of the setting Sun from the eastern sky.) Steve is not sure if the vague grey area we can see between the clouds in the east is in fact the Earth's shadow. But soon he spots a rosy section above it, which may be the Belt of Venus. This soft pink 'belt' is made by scattered sunlight and appears just above the Earth's shadow. So if this reddish patch is indeed the belt, it's another sign that the Sun is setting.

We turn back around to study the clearer sky in the west. Steve draws lines in the air with his hand to mark the regions of colour that are now developing there. Above the reddish tint running along the top of the hills is an intense peachy-golden glow. Then a yellow-white strip of cloud. And beside the cloud, he says, 'You can see there's a hint of greeny blue.'

Greeny blue?

We all perceive the colours and qualities of light differently. That means we're not seeing quite the same twilight. But now Steve draws my attention to it, yes! I can discern a faint stripe of rock-pool green. As a culture, we so often delight in the golds and reds of sunset. After this revelation,

I don't think we celebrate nearly enough the turquoise tints that add nuance to its hues.

That greeny-blue section, Steve continues, will keep spreading up and out. And over the next hour or so we can expect the colours of the sky to separate into more obviously distinct bands around the bright round glow beaming up from the Sun.

Is it possible to tell what point of twilight we're in by the colour of the sky? To an extent, Steve affirms, as long as clouds don't muddle the picture.

He pulls a book from his pocket and presses open the pages to catch the light from the west. The book is a recent edition of *Light and Colour in the Outdoors* by the Flemish solar physicist and astronomer Marcel Minnaert. This hugely popular scientific work was first published in the Netherlands in 1937, around the time the *Nautical Almanac* introduced its three-part model of twilight.[14] While the *Almanac*'s three twilights describe the key changes in visibility in ideal conditions, Minnaert's richly descriptive narrative adds fascinating details.

Steve has *Light and Colour* open on a chart that shows what happens in the sky to the east and west, as Minnaert says, during 'a typical sunset in medium northerly or southerly latitudes on a clear evening'.[15] (That includes this part of southern England, at roughly 50° North.) The extract, reproduced below, sketches four rough snapshots of the western sky as the Sun drops below the horizon (0°), degree by degree. The hard lines between the colours represent soft transitions in the sky.

This chart brings to mind a line in Thomas Hardy's novel *The Woodlanders* (1887). The country person who can't rely on clocks to discern the time of day, his narrator observes, 'sees a thousand successive tints and traits in the landscape'.[16]

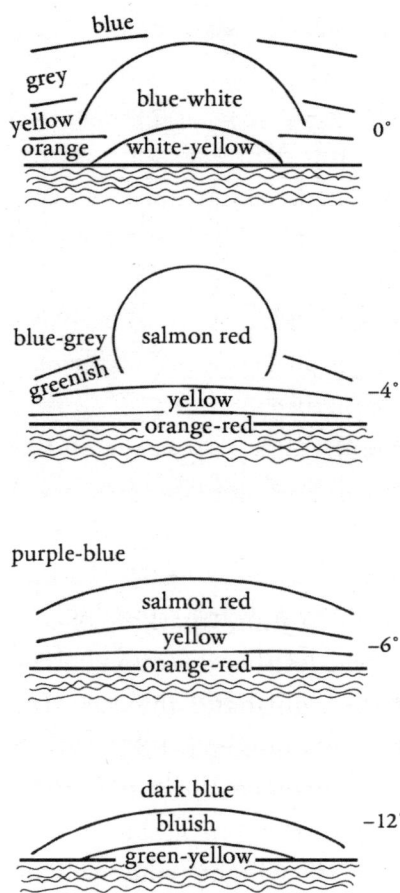

The word 'traits' might mean many of the signs of time we've encountered in this book: shadows, cockcrow, birdsong, cows walking themselves in for milking at the same time every day. Whereas, for me, the word 'tints' most obviously implies the changing colours of the light.

The potential for sensing *time* by the light is even more apparent from Minnaert's second table, below, which tracks major events around sunrise or sunset against the clock.[17]

The first and second columns chart the Sun's altitude above and below the horizon (0°) by roughly how many minutes

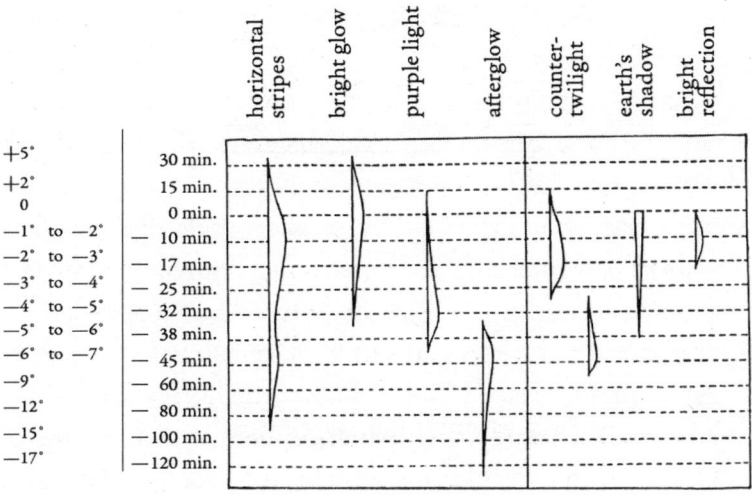

have passed, with '0 min.' marking the moment of sunset. Inside the box, the left-hand section shows what happens in the west, and the right-hand the east. The irregular-shaped bars represent sky phenomena and thicken when they become more intense. For instance, the 'bright glow' in the west peaks at sunset (0°). And the 'purple light' increases after sunset, then fades after civil twilight (when the Sun is 6° below). In actuality, the subtle 'purple light', Minnaert explains, 'radiates colours of a wonderful soft transparency, more pink and salmon-coloured than true "purple".'[18]

Out here in the fields, the reddish tint along the top of the hills seems to have expanded upward and turned rhubarb pink. A band of orange-yellow light has replaced it in the lowest part of the sky. The pale wash of greeny blue has spread out and up.

Without Steve's help, though, I'd find it very hard to identify patterns in this kaleidoscopic sky. It takes practice,

Minnaert writes, to perceive 'the fine ranges of colour, evanescent, tender tints'.[19]

Minnaert is describing how the physicist, not the artist, observes the light. But there is a delightful blurring in his book between the two. Indeed, he seems intent on setting us all alight with the pleasures of exploring the sensory world like the poet-scientists of a former age. In a charming flourish at the start of the book, he enthuses about the nature-lover who 'wanders over the countryside, eyes and ears alert, sensitive to the subtle influences that surround him, inhaling the deeply scented air, aware of every change of temperature, here and there lightly touching a shrub to feel in contact with the things of the earth, a human being supremely conscious of the fullness of life'.[20]

Nor does Minnaert neglect the town-dweller. His assertion that 'even in the noise and clamour of our dark streets, the manifestations of nature remain' is something I've taken to heart.[21] And this is one of the phases in the day when he encourages us to look up. Around half an hour after sunset in the midst of the city, he writes, 'in narrow streets from where no western horizon can be seen', the glow on west-facing 'buildings shows clearly that the purple light is shining'.[22]

One summer evening I thought I caught an especially lovely glimpse of this 'purple' light while walking through my local street market. The stalls had packed up, the street was swept clean, and there was a dreamy mood about the place after the bustle of the day. A hairdresser leaned in a doorway, slowly combing a wig. Shopkeepers sat out chatting. The low drone of sports TV drifted from the Good Intent. When a bright burst of laughter floated down from a balcony, I glanced up to see the top slice of a tower block glowing syrupy yellow. A little while later, when the light was dimming, I saw the top of the block had turned a gorgeous pink.

THE GLOAMING AND THE DIMPSE

My friend Jmeel Allen – the designer and photographer who marvelled with me at the great drift of blue chicory flowers in the park – is someone who looks very carefully at the world. And he shared an intriguing insight about urban skies. Jmeel grew up in the Welsh mountains but lives in London. To his surprise, in some ways he finds sunsets here are more startling. The city's angular shapes, he told me, make a play of contrasts with the flowing light. A square archway opens up to reveal a molten lake of purple cloud. Tower blocks frame the white disc of the rising Moon. The long strips of window on a deteriorating housing block gleam like liquid bands of rose gold.

Jmeel often heads out into the city with his camera at the end of the afternoon. But for so many of us, the constraints of working hours and indoor life mean we tend to miss the drama of the sky. How might we be coaxed over to the window and out into the street to enjoy it more?

In *What Time Is This Place?* (1972), the visionary American urban planner Kevin Lynch lamented how our increasingly indoor existence removes us from signs of time from natural cycles. 'Office and factory buildings, long corridors, and subways are timeless environments, like caves or the deep sea. Light, climate, and visible form are invariant,' he wrote.[23] By 'timeless environments', presumably Lynch meant places where natural signs are absent and the abstract, metered kind of time dominates. He noted, for example, that in underground travel networks 'only clocks measure the empty passage of the hours'.[24]

In response to this lack of variety in urban environments, Lynch proposed introducing 'arrangements that amplify or complement the underlying natural clues of time'. Streetlamps could darken as the night deepens, he suggested. Or they might change with the Moon's phases. And there could be 'reddish lights that magnify the sunset'.[25]

Today there is an 'arrangement', to use Lynch's expression, in the London Underground station for Battersea Power Station that I expect would have delighted him. When you come up from the escalator to the station concourse, you'll see above you a richly coloured frieze running in a horizontal strip for a hundred metres along the wall. This is a permanent artwork, *Sunset, Sunrise, Sunset* (2021), created especially for the station by the artist Alexandre da Cunha.[26] And it represents quite another kind of time from the digital clocks on the station information boards.

The frieze is an abstracted mosaic of a twilight sky made from vertical metal bars. Each bar is coated in one colour from a palette inspired by photographs of sunsets and sunrises in London (from my notebook: coral, shrimp pink, pomegranate, lemonade, pigeon grey, blackcurrant…). The vertical bars are, in fact, triangular prisms. And on my first visit to the station, that suddenly became apparent to me when a shiver ran from one end of the frieze to the other as the bars swivelled in place to reveal a new mosaic of the 'twilight sky'. The movement felt familiar – the same technology is used in rotating billboards – and in this busy mechanical environment, the artwork has a subtle presence. But *Sunset, Sunrise, Sunset* is wonderfully disconcerting if you stop to study it. I waited for the ripple to run through the bars again and discovered the revolution didn't happen with a metronomic beat, but when I wasn't expecting it. Then, beyond the ticket barrier, there's a second kinetic frieze much like it, with a different rhythm.

Sunset, Sunrise, Sunset is more disorienting and ambiguous than Lynch's proposal for reddening streetlamps. As its title implies, the artwork's colours change and return. But they do so in unpredictable, irregular phases without beginning or end. The shape of time they describe feels palpably different from the onward march of identical mathematical units coordinating life on the concourse below. When we're rushing through the station, the strips of rippling colours on the walls might suggest some kind of gateway or threshold to other experiences or ideas of time. At its simplest, *Sunset, Sunrise, Sunset* is a gentle reminder to look up from the screens in our hands and open our senses to the dusk and dawn.

Here in the Somerset fields, Steve and I are losing hope for a pristine evening sky. Soft lozenges of cloud are floating above the western hills like shards of raspberry-pink ice melting in water. Higher up the sky has laced over with drifts of haze. The clouds, in other words, are masking and muddling the full range of what Minnaert described as sunset's 'evanescent, tender tints'. Moreover, the sky will seem redder for longer tonight because the clouds will keep reflecting the rays reaching up from the Sun.

The sequence of sky colours, in short, is not conforming to Minnaert's charts. So my hope that we could trace finer phases of time by their pattern is scuppered.

Despite this hitch in the plan, the watery sunset is fascinating to watch. A short while ago we noticed grey-purple plumes wiggling over the clouds like thick foamy threads, and even Steve wondered for a moment what they could be. Then we realised they were aeroplane contrails broken up into peculiar shapes by the wind and oddly lit by the rays from below.

If I had an aerial view of the countryside just now, I wonder how many farmers I'd see dotted about the fields enjoying the burnt-marshmallow pinks and orange caramels that have begun tinting the clouds. When Steve brought me to this lookout, it felt immediately familiar because often at the end of the busy day, my parents, especially Mum, would lean on a field gate admiring the towers of colourful clouds rising over the Midlands.

I suspect this red-sky-watching ritual goes back countless generations. And if there's an ancient piece of weather lore many of us still know, it's 'Red sky at night, shepherd's delight. Red sky in the morning, shepherd's warning.'[27] The Met Office confirms there is a wisdom to it. In the UK, a red sky in the evening could mean fair weather is coming in from the west, and in the morning that it's passing on to the east.[28]

In the era before Met Office broadcasts, a red sky could be a critical sign of whether a brighter tomorrow is coming over the horizon. In that sense, the colour of the dusk and dawn sky foretells the time – what it will be possible to do, and what opportunities should be seized.

In the fields around us now, the pheasants are crowing with more urgency. A blackbird shrieks an alarm call. Pale mists have gathered in the shallows. The air seems to have gained a deeper bluish tint, particularly down in the dark mass of a distant spinney.

I wonder if this is what photographers call the 'blue hour'. That unofficial term has various definitions. The meteorologist Stephen Corfidi later suggested to me that the blue hour could mean soon after sunset, when the light on the ground starts to lose its redder hues. (This is another effect of different parts of the atmosphere filtering sunbeams as their angle alters through the day.) That change in the colour of the light is why, this evening, the pages of Minnaert's book looked greyer and bluer not long after the Sun sank behind the hills. But some photographers place the blue hour later, starting around now, towards the end of civil twilight. This is when the Sun has dropped so far that the purple light is weakening and the blue light dominates. In ideal conditions, the darkening sky and air take on gorgeous hues of deep blue.

Ragged scraps of cloud are presently drifting overhead and obscuring the celestial blues from my awareness. But what I am alive to is the quality of the light on the ground. In the field and lane now, the air seems distinctly dimmer and duller. Maybe this is the state of 'murky half-light … at the end of the day' that people here in Somerset would call the dimpse or dimpsy.[29]

Nevertheless, in the western sky, the glow is still vast and from time to time the coral and pink undersides of the smudged-charcoal clouds brighten like hot coals. For me, this is the moment when the split becomes most palpable between the gleam above and the gloom below.

'Oh look, Steve!' I say, 'the lettering is going.'

I've been waiting for this. The pages of *Light and Colour* have been growing steadily less legible until it's almost impossible to read the paragraph that explains, when reading 'becomes difficult, the "civil twilight" is over'.[30]

Civil twilight this evening will last around thirty-five to forty minutes. So, by checking his watch, Steve confirms the Sun has now dropped to around 6° or 7° below the horizon. That's when the purple light withdraws, Minnaert says, leaving us with the sensation that the land is darkening fast.[31]

So far, then, the light is dimming to plan. On the other hand, Steve advises, 'You have to be careful making sweeping generalisations, because if you've got cloud close to the horizon, that can modify what you actually see.' The low cloud tonight could reflect the light down to us and give us 'a bit of an extension of the glow'.

In the mid-twentieth century, the *Nautical Almanac* defined the average state of visibility in clear conditions at the end of civil twilight in surprisingly vague terms, at least to this reader. In general, when the Sun has reached 6° below the horizon, 'ordinary outdoor civil operations are difficult without artificial light, although there will still be ample light to make possible large scale operations, requiring outline only'.[32]

When the Almanac Office team explains what this means to non-experts (criminal investigators, sports-event organisers), they use more relatable rules of thumb. Civil twilight is over when the light is so dim you have to give up playing ball games like football or cricket. Or (much like Minnaert says) it ends when you can no longer read a newspaper.

In his popular Italian travelogue of 1769, the great French astronomer Jérôme Lalande used almost the same expression, 'when it starts getting hard to read'.[33] The phrase appears in his explanation (to save the traveller from embarrassment) of just how differently the clock hours are counted in Italy.

In most of Europe the cycle of hours ended and restarted at midnight, meaning the hour bells chimed at the same fixed points every day. In Italy the clock kept twenty-four uniform hours, too, but the cycle ended and started at dusk. The twenty-fourth hour struck *half an hour* after sunset, Lalande wrote, when night is falling and reading is beginning to be difficult. That meant the Italian hours changed timing throughout the year.[34]

In places where you can't dismiss the darkness at the flick of a switch, maybe the moment when you can no longer read by daylight would be a familiar beat in the common rhythm of the day. There is an old expression for evening twilight, the 'dark-hour', which *The Vocabulary of East Anglia* (1830) by Robert Forby defines as the 'interval' between there being enough light to work or read 'and the lighting of candles'. This was 'a time of social domestic chat', for example, 'We will talk over that at the *dark-hour*.'[35] In London, it was said, people called this time 'between lights'.[36]

Lalande was an astronomer, and I've not been able to discover whether clocks were set in Italy by identifying the moment when lettering becomes hard to discern. Either way, his expression suggests this method of marking time goes back at least to the eighteenth century. And maybe it is far older and more widespread. I've come across two very different examples of finding points in twilight by testing the visibility of fine details. There is a Buddhist tradition of starting the day when you can see the three main lines on the palm of your hand, but not yet the finer lines.[37] And in fourteenth-century Paris one guild of (apparently commercially minded) leatherworkers were not to start work in late winter until they could tell between coins from two different cities.[38]

The Italian clock system was not unique in Europe and elsewhere in finding a way to make clocks keep track of the fluctuating time of sunset or dusk, which was so important to the rhythms of life for past generations. As it still is for communities in the UK and around the world.[39]

For Muslims, sunset is a particularly significant time. This is when each new day begins in the Islamic calendar. And it's the time of Maghrib, one of the five daily prayers, whose timings are all measured by the Sun: by the length of shadows, for instance, and the first streaks of light at dawn.[40]

The timing of Maghrib, like the other prayers, changes from day to day and place to place. At British latitudes, sunset palpably moves earlier and later through the seasons. And on the same day, its timing depends on where you are. Today in Taunton, for example, sunset happened around seventeen minutes later than in Nottingham in the Midlands to our north and east. Muslims have a strong sense of the timing of local sunset, the social historian Imad Ahmed told me later. 'It reminds me of how, once upon a time, before Greenwich Mean Time became standard across the land, clocks used to be set according to the local Sun,' he said. 'When you pray according to the Sun, even today you are still tuned into where it is in the sky above you.'

For many centuries, Islamic timekeeping experts have produced astronomically calculated local timetables to give clarity and guidance on when to pray. In the UK, Imad said, many mosques use the raw data provided by HM Nautical Almanac Office to calculate prayer times according to their own definitions.

Traditional practices of observing in person have also continued at least until recently. Imad's father grew up in a rural part of what is now Bangladesh, where his community reckoned time for prayer by checking shadows and the sky. Imad once asked him what people did on cloudy days. His dad replied that they timed the Fajr prayer at dawn by the cockerel crowing. And the Asr prayer in late afternoon by the snake-gourd blossoms starting to open up. This most exquisite flower has white petals, edged with fine nets of feathery threads, that slowly unfurl through the night. 'My dad's account of his life in the village – telling time by animals, plants, the Sun and shadows – really moved me,' Imad said, 'and helped me realise how disconnected I was from all of Mother Nature around me.'

I got in touch with Imad because he's the founder of the New Crescent Society, a network of Moon-watchers across the British Isles. Each new month in the Islamic calendar begins as soon as the new crescent, Hilal, is sighted. This happens when the Moon reappears (after disappearing for a couple of days) as a slender arc of light just after sunset, near to where the Sun sank.

Until now, Imad explained, British Muslim communities have relied on sightings from other countries. But the new crescent Moon isn't visible everywhere around the globe simultaneously or on the same evening, which means communities begin the Holy Month of Ramadan, for example, or celebrate Eid on different dates. Imad started the society to see if it was possible to form a unifying religious calendar for Muslims in this country by encouraging people to go out to look for the crescent and share their sightings.

The obvious challenge is the weather. But after several years of monthly observations, the society is finding that when the sky is overcast in, say, the Cairngorms, Cardiff and Cornwall, it is likely to be clear somewhere else. Imad is now confident that the new crescent Moon can be seen by someone in the network across the land each month. If you're in Cambridge, everyone is welcome (Muslim or non-Muslim) at the monthly sighting events Imad leads from Cambridge Castle.

The tagline to the New Crescent Society is 'connecting the community to the cosmos'. The aim, Imad told me, is 'to bring back the human experience to the lunar calendar – to connect with the Moon'. After organising dozens of crescent-sighting events, he has seen how people 'find wonder and joy in connecting with nature'. Elders in the Muslim community have been moved to tears because the experience brought back memories of going up onto the roof as a child to watch for the Moon. 'This was part of life one or two generations back,' he said, 'and now it's returning again.'

'When we look for the crescent,' Imad reflected, 'it's up to everyone to make our own meaning – and it changes throughout life.' One constant for him has been that the new crescent Moon represents renewal and hope. 'After the darkness,' he said, 'the light returns.'

In the sky above Steve and me, the Moon is broad and waxing. And I've noticed something else beside it.

'Steve,' I say, 'the first star came out a moment ago. There it is – to the right of the Moon.'

This bright star is Jupiter, he confirms. I realise he must have had his eye on it for quite a while. If I had known what I was looking for, and begun scanning the southeastern sky between the clouds, I might have spotted it too.

The planet Jupiter is the fourth brightest celestial object, after the Sun and Moon. The third brightest object, the planet Venus, will appear in a few hours, as the night draws towards dawn.

Venus is the morning star. Or Venus is the evening star, depending on the season.[41] Later this year this gloriously distinctive planet will disappear for a while then start showing up close to sunset. According to the early medieval English Christian monk Bede, the time of day called *uesperum* (eventide) is 'when the star of that name' Vesper (Venus) 'appears'.[42]

The first points of light in the dimming sky give signs of the time in more than one tradition. There is a living practice, for example, in Ukraine to wait for the first star before beginning the Holy Supper on Christmas Eve. Children are usually tasked with watching for it, and it's easy to imagine their excitement. But I've heard that since Russia's invasion of Ukraine, festivities have been much sadder and more subdued.

THE GLOAMING AND THE DIMPSE

Here in the Somerset countryside, we're moving into nautical twilight, and that means we can expect the brightest among the 'fixed' stars to have appeared. Sure enough, when we look all around between the clouds, there are two or three more bright sparks in the ultramarine sky.

Steve turns to the west to show me what's happening now. The golden glow from the Sun has sunk low and is losing its vibrancy. But the filigree silhouettes of treetops in the foreground still bristle above the hazy domes of the distant hills.

When we reach the end of nautical twilight, Steve says, we won't be able to tell exactly where the horizon is. As this moment draws closer, the line that splits the Earth from the celestial vault is becoming blurrier. To show me how visibility is already fading, Steve holds his hand in front of the fields. I can't really see it until he waves his hand higher, against the brightest part of the sky.

As we move through nautical twilight, we're pressing closer and closer to the dark shore of night.

Steve and I lean on the gate in quiet contemplation, and my mind drifts. In the growing gloom, there's a sense of stillness around us and I notice the birds have fallen silent now. I guess this is the time of evening meant by the Old English word *cwyld-seten*, whose associations imply the quelling or the hushing of the day.[43]

'You can see there's a kind of whitening going on,' Steve says after a while.

I wrangle my attention back to the sky and realise the vivid reds have almost faded, and the weak gleam of light over the western hills is quelling too.

But Steve isn't only noticing how the colours are fading from the sky.

He turns around to the lane. 'You know the car is blue,' he says, 'but you can't tell from just looking at it now.' And in front of us now, it's much harder to tell this is a grassy field.

I tussle inwardly to forget what colours the dark-blue car and the green field *should* be. Then, if I didn't know otherwise, the car is definitely black and the grass paddock could be a pale mud yard.

The twilight-colour clock, as I'm calling it, has revealed its most obvious sign: the world around us has turned monochrome. And I wasn't really noticing.

'It's actually our eyes,' Steve says, that aren't 'picking the colours up any more'.

What I vaguely believed was happening only outwardly, in other words, is happening within me too.

In daylight, our eyes rely on the cone cells clustered at the centre of the retina. They give us the ability to perceive detail and colour. Very simply put, as the light dims, the cones gradually stop responding and the rod cells around the edge of the retina slowly take over. While the rods are much more sensitive to the light, they reveal the world in lower resolution and without colour. That's why, if we were lucky enough to see a rainbow cast by moonlight, it would appear as a white arc.

Over the ages, intricate changes in colours in twilight were no doubt noticed by all kinds of people with refined sensory awareness. And evidently the appearance of colours could be observed to mark time. According to an authority in Mishnah Berakhot (a collection of Jewish oral traditions composed around 200 CE), it is time to say the Shema prayer in the morning when you can tell between blue and green.[44]

One of the most intriguing and subtle aspects of the shift from day to night is how colours don't fade uniformly

THE GLOAMING AND THE DIMPSE

at dusk. They appear to shuffle hierarchy, with the brightest of daytime colours dropping back as if to let more reticent colours come forward overnight.

In the early nineteenth century, the Czech physiologist Jan Evangelista Purkyně walked out in the spring countryside and noticed curious shifts in the colours of flowers after sunset.[45] Most famously, in his analysis that followed, he observed how the reds that appear brightest in daylight are the darkest colours in dim light, while blues seem brighter. Although he wasn't the first to study this shift, it is known to us now as the Purkinje Effect.[46]

It happens as the light dims and our vision starts relying more on rods, which are most sensitive to the blue end of the light spectrum – and less on cones, whose sensitivity peaks near the red end. In between the two extremes of bright light and darkness, both the cones and rods are active together. That means we can see colours, but we perceive them differently, because the balance is tipped more towards the rods' sensitivity to the blue-greens. Then when the light is so weak that the cones are inactive and the rods dominate, the blue-greens still look brighter than the reds, even though both now appear to be shades of grey. That must be why the grass in front of Steve and me looks both light and colourless.

After he shows me this transformation, I keep watching out for what appears to happen to colours in low light. On an evening walk, I notice a red 'SLOW' sign now looks burgundy and seems to recede against the grass behind it. At home I see for the first time how the colours of a poster look in the near-dark. The vivid reds and pinks are almost black while the duck-egg blue appears strangely pale.

These autonomous changes in our senses, enabling us to see more in darkness, are self-protective and magical – and, I admit, they feel a little eerie.

The word 'day', by one theory, might be related to very old terms to do with heat.[47] That makes tangible sense – we feel the difference between day and night not just through light and colour but temperature too. As Gísli will tell me when we're talking about the darkness returning to the bright subarctic summer night, twilight is a 'multi-sensory experience'. Right here and now, as twilight deepens, I'm starting to feel the chill on my skin and am newly aware of being able to see my breath.

In the past, I guess the dankness in the atmosphere would be felt as a sign that the 'cold sicknesse', 'the infected ayre', 'the vnwholesome mists' are rising. Those are the fearful names that the Elizabethan writer Nicholas Breton gave to the polluted vapour that people believed hung in the air at night.[48]

I pull my woolly hat down over my ears and zip my coat to my chin. Steve, in contrast, is wearing an impressively thin jumper. 'I dare say my nose is a bit chilly,' he says. But he melts in the heat, he explains, and this is his kind of weather. We lurch into silence at the sound of a high-pitched squeal from the fields. After a moment we figure it's just the whistle from a dog walker.

In actuality, this is not a scary place to be. We're close to the village and there are people about. Yet in the dull dimpse, the grey hedge looks lumpy and weird. The sudden movement of a tree branch is slightly disconcerting. It's getting easier to believe this is the time when ghouls and goblins are starting abroad, and witches will be riding about on wolves.[49] As the darkness spreads overhead, leaving a smaller and smaller aperture of light in the southwest, from a certain frame of mind, it's like being swallowed by a great nocturnal beast.

Soon enough, a couple of dog walkers pass behind us on the other side of the lane. Their figures are a loose collage of vaguely distinctive shapes. If I had to identify this pair in a police line-up, I don't think I could do it.

In our brightly lit habitats, we no longer attend to the moment when people become indistinguishable at dusk or identifiable again in daylight. Yet there are methods and expressions from around the world that mark time by this critical shift. A fourth-century account by Egeria, a Hispano-Roman Christian pilgrim, refers to the hour when people start to be able to recognise each other.[50] In a fourteenth-century regulation for Parisian fullers, work should start as soon as you can recognise someone in the street.[51] 'When a man's face can be known' is an expression for a time of morning twilight in the Society Islands in the South Pacific recorded by a missionary in the early nineteenth century.[52] An old Irish expression for daybreak among country people has been translated as 'light when one cannot distinguish between a man and a bush'.[53]

Some of these phrases recall for me that unnerving term for twilight, 'between dog and wolf', *entre chien et loup*. This is the time of day when the sensory world becomes more uncertain.

And now that strange sound is back.

'I heard something there, oh yes,' Steve says.

The squeals from the fields have gathered pace until, with a little relief, I recognise them to be the calls of owls.

Some people get spooked when out observing in darkness, Steve reflects. But he's used to the calls of crepuscular creatures, the rustling of unknown entities in the hedge, the flitting of little bodies in the air. Leaning on this gate under the stars, watching and listening to the night drawing in, is a good place to be.

It will be a while yet, Minnaert predicts, before the last sliver of light hanging on in the west fades away.[54] That will happen when the Sun reaches about 17° below the horizon. We're not at that stage, but it feels imminent to me. Steve agrees – from now on for the rest of twilight, the quantity of light reaching us on the ground won't alter very much. The average person, he reckons, 'would say it's dark at the end of nautical twilight'.

Over the last couple of hours, the transformation of the world has been surprisingly rapid as each shift in the light changed what we saw, felt or heard. Low sunbeams turned golden and shone through the trees. Shadows stretched and dissolved. The colours of the sky flowed in bands of red and yellow and tints of green. Mists rose in the valleys. The air turned bluish. The first stars lit up. Letters blurred. Silhouettes became blunter. Colours faded. The horizon dissolved. More stars shone between the clouds. The air felt dank. Owls called. And now the beat of change is slowing to a stop.

In the course of this one evening, I've grasped nothing like Steve's sensitivity to sunset's phases. But thanks to him, twilight is no longer a chaos of coloured light and shadow to me. I'm more alive both to the patterns that repeat each twilight and to the unique details that will never return in quite the same way. Even on busy days when I catch sight of a fragment of the sunset sky, it's an uplifting joy to guess from its hues what broad phase of twilight we're in. And like watching for flowers folding, there's a quietly defiant pleasure in honouring this great surge of events in living rhythms that the clock on the wall simply ignores. At the end of the day, to devote a little time to luxuriating in twilight skies is to celebrate, in Minnaert's words, 'the fullness of life'.

The cold has set my teeth chattering, and Steve and I decide to head back indoors. What was soft mud now crunches with frost under our boots. Before we leave, I take one last look to absorb the scene.

All the colours of the world seem to have reduced to two red specks twinkling in the dark sky: bright Aldebaran in Taurus and Betelgeuse in Orion.[55] Between the clouds, fainter stars are beginning to sparkle too. The night is nearly upon us.

Now it's time to head deeper into the darkness to track the phases of night by the motion of the stars.

7

Star Clocks

Marking time by the motion of the night sky

The Baker Street Irregular Astronomers are an informal monthly gathering of night-sky enthusiasts with special permission to meet after dark in Regent's Park. The Irregulars (for short) collect at the Hub, the glass pavilion which glows at night like a small round spaceship hovering above the game pitches. The meetings are hosted by a (self-dubbed) Unofficial Force of friendly astronomy specialists and photographers. Everyone is welcome, including novices like me. And this is my first evening as an Irregular.

It's mid-October 2023, around the time Steve Bell and I are hatching a plan to watch twilight unfold. The nights are tangibly beginning to overpower the days, which makes this the start of the best season to go stargazing. Regent's Park is a relatively dark spot in central London's vast lake of light, and I'm told you can get a pretty good view from here of the brightest lights in the nocturnal sky.

Until a few minutes ago, however, the main object in our sights was thoroughly obscured. 'I've found the BT Tower Nebula,' someone joked, looking up from a telescope. This tall tower displays advertising messages over central London and is emitting an intense glare of pink, blue and red light that was far outshining the planet Jupiter, the brightest celestial body in the sky tonight, as it rose from the horizon.

While we waited outside on the Hub terrace for Jupiter to triumph over the urban glow, the Irregulars stood about, chatting affably and checking what they could see from time to time. I noticed how the people with telescopes first squinted at the sky before they lowered their eye to the eyepiece. And when someone eventually pointed to the slightest hint of a gleam from Jupiter, and others agreed there was something there, I was perplexed that I couldn't see it. Obviously, shielding my eye from bright light is an essential first step if I want to see more stars. But now I begin to understand, from the patient postures of the people around me, that I need to slow right down, stay still, and devote full attention to the sensation of the tiniest glimmer in my peripheral vision. Or rather, as astronomers say, *averted* vision.

In the special conditions of a gallery or a cinema, I know how to peer intently. And after an adulthood of rushing through my everyday surroundings in a state of distraction, through the adventures of this book I've become much more attentive to petals folding, shadows lengthening, and the rising and fading of sky colours. Nor am I utterly unfamiliar with the stars – I can locate the North Star and the constellations of Orion and the Plough (Ursa Major or the Big Dipper). Yet, for me, trying to tell time by the stars presents a new level of complexity and subtlety unmatched even by the phases of twilight. I am daunted and excited.

Friendly members of the Unofficial Force are on the terrace outside the Hub welcoming newcomers, answering questions, helping orient telescopes. Over the coming months, the two I'll learn from most are Eric Emms, an amateur astronomer

THE FULLNESS OF TIME

(whose fascination was ignited, he says, by his parents' gift of a toy telescope in the Apollo era), and Nick Joannou, an astronomer and tech specialist with especially vast knowledge. With relief, I soon learn how enthusiastically the Force welcomes any question about the sky, no matter how simple.

Tonight we're all invited to explore Jupiter's splendidly marbled disc through the telescopes lined up on the terrace. But I'm here to seek help with my own very particular ambition. This winter I want to discover through experience something of how ordinary people were able to look at the sky without tools or maps and know how far the night has progressed.

Imagine you've just woken up. You reach for your phone and see it's not yet 3 a.m. Checking the time in the darkness is a comforting reflex, like touching the side of the pool before diving under again. You close your eyes and settle into your pillow. Then a doubt bubbles up. Maybe this is the night the clocks go back.

Not that long ago you had to remember to alter your watch and thermostat and every other object with a timepiece in your life. But these days, more and more of your devices reset themselves. So, while you were sleeping, perhaps the clocks paused and fell behind the Sun. You might have gained an hour without knowing. Lying here in bed, there's no obvious sign to tell for sure unless you check online. In the stillness and darkness, time seems illegible. The ambiguity of the hour, the doubtfulness of the clock, is disorienting.[1]

We're so tied to clocks that on those occasions when we become aware of their autonomy and our dependence, it's unsettling. What *is* time for us when you take the clock away? How do we orient ourselves and gain a stable foothold in time? These questions are most pressing, I think, in the darkness.

The kind of time-vertigo I'm describing may be special to our age of machine-given, super-precise hours. After the shadows have dissolved and the Sun has sunk, when day's-eyes have closed and the street is silent, I've no doubt that past generations were more experienced at distinguishing broad phases of darkness by the crowing of the rooster and the whistling of the early songbirds. I had assumed, though, that the night lacks a concrete sign as simple and clear as the shortest shadow. But it turns out that's not so if you know how to read the sky.

A few months before I found my way to the Irregulars, my intrigue was sparked by a detail in *Far from the Madding Crowd* (1874), the novel by Thomas Hardy. The beleaguered (yet, unusually for Hardy, not ultimately doomed) main protagonist, Gabriel Oak, was a shepherd in possession of a faulty watch. It either ran 'too fast or not at all' and you could never be sure if the hour hand had slipped on its pivot. Yet this resourceful farmer wasn't disarmed by his faithless timekeeper. Not only did he sneak glimpses of clocks through windows, but he constantly checked the Sun and stars.

We've already seen how people loosely reckoned the time of day from the angle or length of shadows, or the Sun passing a landmark. Yet how Gabriel could figure out the time by the stars was a riddle that set me doing a little detective work.

Close to midnight on the eve of the shortest day, the narrator of *Far from the Madding Crowd* leads us out onto a windy hillside where Gabriel was tending to his ewes and lambs. The sky was glittering with stars, and 'the Bear had swung round [the North Star] outwardly to the east, till he was now at a right angle with the meridian'.

First, by 'the Bear', I take Hardy to mean the seven brightest stars in the constellation of the Great Bear (Ursa Major), which today we often call the Plough or Big Dipper and latterly, Charles's Wain.[2] It was midnight on midwinter solstice and the 'Bear' was east of the North Star.

Then two months later Gabriel fell asleep in a stationary hay wagon and awoke disoriented to find it moving at a fast clip. (The wagon, we later discover, was carrying him to his future.) He peered up from the hay at the stars, and especially at the celestial wagon, Ursa Major. 'Charles's Wain was getting towards a right angle with the Pole star, and Gabriel concluded that it must be about nine o'clock.' Even in his distracted state he read the time by the stars 'without any positive effort'.

Hardy seems to have given us the same image twice: Ursa Major is to the east of Polaris (the Pole or North Star).[3] It's almost as if we're receiving a lesson in telling time by the stars – although it was not obvious to me how to make sense of it. Why would the stars circling Polaris be in the same place in the sky at midnight at midwinter solstice and at 9 p.m. in February?

These days you can download powerful night-sky apps that instantly chart any part of the sky you care to point your smartphone at. You can run the clock function to see a simulation of the stars moving through the hours of any night in any place. But I wonder how nineteenth-century laypeople would make sense of the motion of the stars.

It's tempting to picture Hardy writing about Gabriel telling time while leafing through *Half-Hours with the Stars* (1869) or perhaps *A New Star Atlas* (c. 1872) – two handsome introductions to the night sky by the popular British astronomer Richard Proctor.[4] Hardy is thought likely to have been

STAR CLOCKS

influenced at points in his novels by Proctor's other, more complex and profound works.[5] It's not inconceivable he'd seen Proctor's guides for the novice stargazer too. In any case, I can't help but imagine some of the first readers of *Far from the Madding Crowd* puzzling over Proctor's maps to figure out how Gabriel told time by the stars.

Proctor aimed (to quote his subtitle to *Half-Hours*) to offer 'a plain and easy guide' to the positions of the main star groups at mid-latitudes in the northern hemisphere 'night after night throughout the year'. From our point of view on Earth, the 'fixed' stars don't change position in relation to each other (unlike the Sun, Moon and planets). But as a whole, the canopy of fixed stars is always turning around us. In the northern hemisphere, the only star that hardly appears to move is the North Star (Polaris or the Pole Star) which sits more or less at the celestial north pole. This surprisingly weak spark in a lonely patch of sky stays pinned almost to the same spot while every other star wheels around it. The North Star is the hub of the night sky. Or, as Proctor would later call it, the 'axle-end of the great star-dome'.[6]

'It is worth noticing,' Proctor wrote in *Half-Hours*, that the Guardians 'form no bad time-piece when used with the aid' of his maps.[7] (The Guardian stars are Kochab and Pherkad, two of the North Star's closest companions.) In a clear, dark sky, the Guardians can be seen revolving around the North Star once in almost twenty-four hours. Indeed, all the fixed stars seem to move through the same path and return to the same place in the sky a little less than four minutes earlier each night.

The hour hand on this celestial timepiece – this 'star clock' – continually runs ahead of our earthly clock. But unlike the doubtful mechanism in Gabriel's pocket or our self-setting clocks, the stars never slip on their pivot but stay visibly wheeling around the North Star at the same faithful rate.

The hands of this star clock don't turn as I expected, though. I've become very familiar with how the Sun appears to move 'clockwise' through the sky in the northern hemisphere. Yet the Guardians 'revolve round the pole', Proctor wrote, 'in a direction contrary to that of a clock's hands'.[8] Which prompts my first question for the Unofficial Force, now that I've joined the Irregulars.

Why, I ask Eric, do the stars turn in the opposite direction to the Sun?

They don't, he says, and turns to look at the illuminated mass of central London. If we're lucky with the weather later tonight, we'll be able to watch the bright constellation of Orion rising over east London. It will arc up over the southern horizon, reach its peak above Baker Street due south of us, and sink down into west London.[9] Unlike Orion, the Plough is always to be seen circling the North Star without ever dropping out of sight on a clear night at our latitude. (For us, that means, it's 'circumpolar'.) But the Plough still wheels up from the east and down to the west. The Plough and its fellow circumpolar stars only seem to turn *anti*clockwise because, when you're looking at them, you're facing the Pole Star in the north with your back to the Sun in the south. The clearest way to think of it, Eric says, is that all stars appear to rise from the eastern horizon and set over the western horizon. That's the constant direction of the great star-dome.

The Guardian stars had been used to calculate the hours for centuries by those seeking to find time at night.[10] But I just want to gain a broad sense of the time from the stars that are most often visible over my patch of London, like Gabriel glancing up from the wagon. Compared with the North Star's two Guardians, the seven bright stars in the huge Plough may give a blunter measure of time, but collectively, they're easier to locate on nights of haze or cloud.

The diagram below reduces one of Proctor's even simpler maps (which he published after Hardy's novel) to show just the North Star and the Plough at four times of day and night at winter solstice.[11] That's when we found Gabriel tending the lambs and ewes on the hill at midnight.

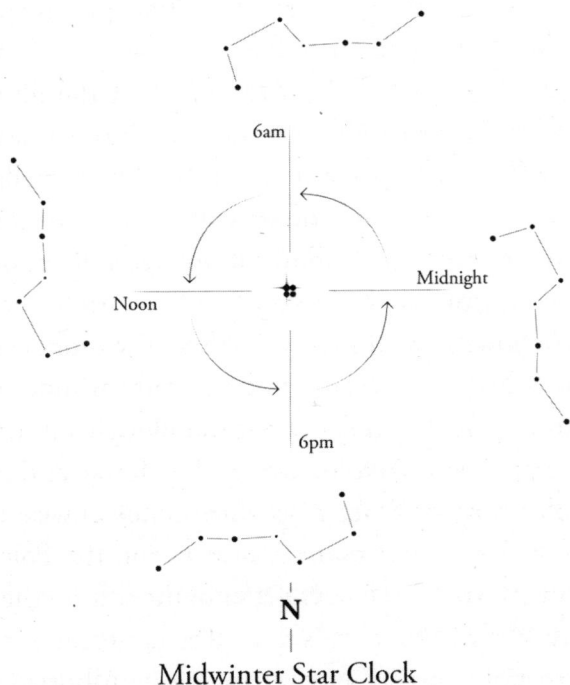

Midwinter Star Clock

The northern horizon is at the bottom of the map. (The gleam on the map marks the North Star's position at midnight in midwinter on its tight circuit around the celestial pole. How high the North Star sits in the sky depends on your latitude. In Reykjavík it's nearly overhead. In Nairobi it's close to the horizon. To find the North Star, look for the Plough or Big Dipper and follow the line given by the two Pointer stars on its outer scoop.) Importantly, this star 'clock' shows local apparent time, meaning midday happens when the Sun appears at its highest point in the local sky and midnight at its lowest point. In London, local apparent time more or less coincides with standard clock time in winter. But when the clocks spring forward to British Summer Time, the stars tell me it's about midnight, and my watch says 1 a.m. Wherever you are, it's worth checking the difference between local apparent time and standard time.

The stars run faster than the twenty-four-hour clock, gaining about an hour every fifteen days. So, after three months, the Great Bear's midnight position will have moved on by about six hours to the next quarter of the sky. Once we know where the Bear sits at midnight in one quarter – in this case, to the right (east) of the North Star at winter solstice – we can figure out its midnight position in the other quarters of the year, as pictured below. Then if we know the position of the Great Bear at midnight on a particular date, we can work out roughly where it will be at the other hours that same night. (This is easier to do if we imagine the circle around the North Star is a twenty-four-hour dial. Through the months, the midnight mark on this dial gradually slides anticlockwise, and all the other hours move with it.)

For Gabriel to have known it was around nine o'clock in February 'without any positive effort', I figure he must have had some kind of system like this.

Gabriel stands in a long tradition of shepherds telling time by the stars – on the printed page at least. In *Le Calendrier des Bergers (Shepherd's Calendar)*, produced in Paris in 1491, instructions for reckoning equal hours by the circumpolar

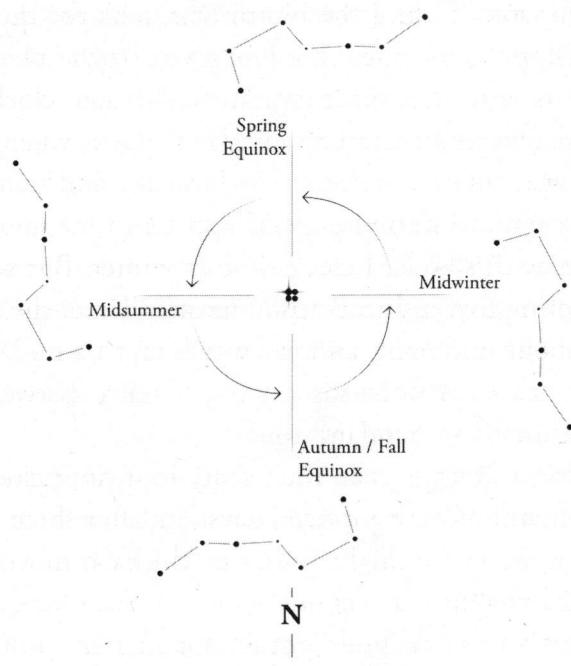

Midnight through the Seasons

stars are accompanied by an illustration of a tired-looking shepherd holding up a plumbline to the sky.[12]

Actual shepherds may not have learned the method from books,[13] or measured the hours by the stars so precisely. But basic principles of tracking time by the Great Bear and other stars may have been familiar, especially to rural people, as a way of navigating the darkness and organising the changing rhythm of the day through the seasons.[14]

During the long, cold subarctic winter, recalled one Icelander born in 1884, the community would come together at the end of the day for the *kveldvaka*, the 'evening wake', in a house warmed by the presence of the cows as well as the fire. People would settle down to work – to spin, weave, knit, braid or mend – while someone read aloud a story or poem.

THE FULLNESS OF TIME

The reader would lean close to the oil lamp and try to be heard above the whirring and rattling of tools. The vigil was long, and clocks were rare. So, people would watch for the Seven Stars, the Pleiades cluster, moving over the daymark for the midday Sun, or one of the other daymarks (we'll come back to this later). The Karlsvagninn (Karl's Wagon or Ursa Major) was another sign of the time. And if they couldn't see the stars, he wrote enigmatically, 'common sense had to be used'.[15]

There is a vivid expression in a Gaelic-English dictionary that hints that marking time by Ursa Major was a part of everyday life in Scotland too. It was said of a visitor who stayed for a considerable time that 'he turned the plough upside down': '*Chuir e car sa chrann*'.[16] (That implies about twelve hours by the clock.) What an elegant way to acknowledge a guest's endurance.

It takes a great deal of mental effort, I find, to figure out where the Plough should be at a particular time, compared with checking a clock. But I can't assume this struggle is inevitable. The archaeologist Fabio Silva is one of the generous experts who has helped me with this project. As he remarked to me, if you lived under a dark night sky, and it was part of your way of life from childhood to mark time by the firmament, you would gain a deep, intuitive sense of the speed and motion of the stars over your home landscape.

The impression Hardy gives in *Far from the Madding Crowd* is that in southern England in the nineteenth century, Gabriel was not unique among country people in marking time by celestial events. But he had particularly sophisticated skills. After his fellow villagers triggered his righteous ire, they tried to pacify him by sharing his reputation for being a 'clever man'. 'We hear that ye can tell the time as well by the stars as we can by the sun and moon, shepherd.'

STAR CLOCKS

Gabriel's enviable skill is why – a few days after meeting the Irregulars and gaining some basic tips – I'm standing outside our flat squinting up at the sky. When a neighbour passes by with an enquiring glance, I explain I'm looking for stars. 'You need to get out of London for that, don't you?' Good point. What hope is there for telling time by the subtle glint of stars while standing between tall buildings in a brightly lit carpark under an intense skyglow?

Over the last century, the spread of bright urban illumination has liberated the night and expanded our sensory experience. I am grateful – this could be a genuinely terrifying place without good street lighting. But the nocturnal glare now leaks far beyond where it's needed or desired. Increasingly we're bleaching away the splendid sky-worlds we used to see nightly by the naked eye.

The Baker Street Irregular Astronomers club was set up, I'm told, 'to bust the myth you can't go stargazing in the city'. Yet my home street is far less protected from light pollution than Regent's Park. Maybe I should hunt for a corner of darkness and wait half an hour or so for my eyes to become more sensitive to low light. But that's not ideal for my ambition to read the star clock 'without any positive effort', like Gabriel Oak.

Happily, it turns out, even here it only takes a few moments for my eyes to grow accustomed enough to see the Plough – and sometimes even one of the Guardians. From now on, for the next few weeks, I research methods for telling time by the stars, and go out on every clear evening to watch the sky from the carpark. (The same surprisingly rewarding carpark, that is, where I've listened to the dawn chorus in spring and peered at the goat's beard – Jack-go-to-bed-at-noon – in summer.)

One night Rosie joins me, and we stand in stillness for a while, our stances focused upward beyond our immediate present. We pick out a few stars sparkling among the racing drifts of cloud. 'All this beauty is with us all the time,' she says. 'We only have to notice.'

I'm excited to show Rosie how I can estimate the time by the stars. We start by locating the North Star, where it always is, above one of the chimney pots on the building opposite. The Plough is low on the horizon and hidden by the building. But I've learned this isn't a terrible problem for my approximate clock. The distinctive w shape of the constellation of Cassiopeia is shining above the rooftops. It's among the circumpolar stars that continually wheel around the North Star, and it sits more or less opposite the Plough. That means, if we can see Cassiopeia, we can estimate where the Plough must be. So now my star clock has two big blunt hour hands. And by the location of the w above the rooftops – I show Rosie triumphantly – we know the time is very, very roughly 9 p.m.

'Heigh-ho!' calls a carrier in one of Shakespeare's plays as he bursts into the yard of a Rochester inn with lantern in hand, 'an it be not four by the day, I'll be hanged: Charles' wain is over the new chimney'.[17] The Wain or Plough is sitting over a particular feature on the roof, and by that he guesses it's four in the morning. For a short while, I feel a little like Shakespeare's carrier, telling time by where the Plough (or Cassiopeia) sits above the skyline.

But each night the star-wheel races a little further ahead, and after a month of unrelenting cloud I'm lost again. How does Gabriel remember where the Bear is at nine o'clock in February without looking at a chart? I need to lodge its seasonal positions in my memory. And for that, I borrow an old idea from sailors.

In 1483, not long before the shepherd's calendar began to be produced in Paris, an important manual was published for sailors in northwestern Europe, *Le grant routtier*, by the French mariner Pierre Garcie. His guide includes a visual memory aid, pictured above, whose applications include reckoning time by the stars.[18]

Garcie's method for calculating time involves imagining the North Star sits at the centre of the circle, like a jewel shining from the figure's middle. His head and limbs divide the sky into quarters, and the Guardian stars turn around him like a clock hand. But I'm not going to attempt to make Garcie's sixteenth-century system work. For my rudimentary purposes, I'm simply adopting the general idea of a human figure in the stars as a way to remember how to tell time by the Plough.

When I look up from the carpark now, I picture the figure floating over the chimney pots with Polaris sparkling at their navel and their feet pointing to the northernmost point on my horizon. I imagine the figure gesturing to where the Plough is in the middle of the night at key moments in the year. And I mirror the movement.

Point your arm straight up at spring equinox. Stick your left arm out to summer solstice. Point your toe straight down to autumn equinox. Point your right arm out to winter solstice.

I wonder if sailors ever did something like this. Would they stand or lie on deck at night and wheel their limbs to get their own star clock under their skin? When I perform this little jig to wind the cycle of starry hours into my body, the memory does stick. Though, if you were to test me, you'd find I'm much better at estimating the time by the Plough during those four seasonal pivots in the year than the months in between.

This basic lesson in star time has been for me the key to deciphering other patterns in the night sky. I'm beginning to devote far fuller attention to the great star-dome. The 'roll of the world eastward' against the stars is becoming simpler to discern and its 'epic' poetry easier to feel.[19] As for my time-telling mission, I'm feeling pretty satisfied by the strides I've made since the last meeting at the Hub and the one coming up this Monday.

Or rather, it's Moonday for the Irregulars. The Unofficial Force chose this date for the November meeting because it's in the phase of the month[20] when sunbeams light the lunar surface from the side, throwing its mountains and craters into sharp relief.

But when I walk into the dark park in safe company, the skyline opens up before us to reveal the heavy banks of cloud that have gathered over London, reflecting the city back at itself in queasy shades of puce and yellow.

What a murky, skyless winter this is turning out to be. While we wait inside the Hub in waning hope of exploring the surface of the Moon by telescope, Nick of the Unofficial Force offers to improvise a talk on lunar themes. Buoyed along by his insights, the conversations roll around the room: whether a short story by the great astronomer Johannes Kepler, *Somnium* (1634) – a dream fantasy about journeying to the Moon – could have inflamed accusations of witchcraft against his mother;[21] how the technical evidence proves the footage from the first Moon landing isn't fake; how it's sometimes assumed the Moon is universally female but is a masculine entity in some cultures, including Norse mythology.

I do have a question about the Moon, prompted by Hardy's villagers saying they can tell time 'by the sun and moon'.[22] We've explored in earlier chapters old ways of reckoning time by the Sun. But how could the villagers tell time by the Moon?

There is at least one phase of the lunar month, I'll learn, when it would not be so tricky to get a fix on the time of night by the location of the Moon. When the Moon is full, it sits opposite the Sun and mirrors its movement, rising at around sunset and setting at around sunrise. But for the rest of the month, the pattern is more complex.[23]

There were various old methods of reckoning time at night by the Moon: Bede provides one for monks.[24] And maybe in Hardy's day some people knew how to reckon the time from the lunar tables given in popular almanacs. That possibility was suggested to me later by the historian Anne

Lawrence-Mathers, who pointed out that ale houses would pin up 'poster' (broadsheet) versions of the almanacs for customers. It's a delightful thought – a tempting bit of fan fiction – to picture Hardy's villagers checking the poster and the Moon to decide if it's time to weave their way home from a tavern.

Right now, though, the challenge of learning to calculate time by the Moon feels too overwhelming. And anyway, I'm brimming with questions about a new set of stars.

On a recent trip to the Cotswolds, the clouds broke to reveal one of the most brilliant skies I've ever seen. For two hours the road had cut a tunnel of light through the blank darkness, and we were a gang of old friends in a world of our own. We pulled up at the cottage in a tiny valley, the engine stopped, and the lamps went off. We tumbled from our warm metal bubble into the cold midnight air, and our sensory 'antennae' raced out into the darkness. 'What's *that*?' my friend Mike called, and pointed to an unreal sight: a cluster of luminous blue orbs which (in my memory at least) were glowing inside a faint haze of spectral light. Close below to the east was a pattern of white stars in a v shape with one red point. And streaming over our heads was a glistening river of milky light.

Would these strange lights have been the Pleiades star cluster and the Milky Way? Yes, indeed, a fellow Irregular had confirmed. The river is the Milky Way. The blue stars are the Pleiades. And that v of white stars is the Hyades cluster in Taurus the bull. The red star is Aldebaran, the 'eye of the bull'. Its name, Aldebaran, comes from the Arabic for 'the follower'.

By serendipity, this close encounter with the stars in the dark valley at midnight is a great fillip to my quest to tell time by the stars like Gabriel Oak.

STAR CLOCKS

On that midwinter night when Gabriel was tending to his flock, he slept briefly and awoke at the sound of a lamb bleating. After finding his watch was faulty again, he went out on the hill to check the time. Instead of turning to the North Star, first he looked south to the stars near the 'restless Pleiades', and then to the rest of the sky, and reckoned it was one o'clock. (Presumably he would have known where these stars would be from comparing his watch with their progress on recent nights.)[25] When I first read this, I couldn't visualise the sky Hardy was describing, let alone fathom how anyone could tell time by it. But now the pieces are gradually falling into place.

The first thing I learn about the Pleiades, or Seven Sisters, is that they're famous around the world as a sign of the season, because for a while each year they seem to leave the sky for a predictable period. I ask Eric why that is. He picks up a coffee cup to model what appears to happen in the sky from our point of view on Earth. The Sun not only circles around us daily but travels through the constellations of the zodiac over the course of the year. When we look up from Earth, he says, certain stars like the Pleiades appear to be very close to the annual path of the Sun. So, imagine this cup is the Sun. The bottle on the bar over there is the Pleiades. When the cup is between us and the bottle, we can't see the bottle. Just so, the Pleiades are veiled by the Sun. The Pleiades haven't actually left the sky, they're just rising and setting in the daytime. When the cup moves on, the bottle starts coming back into view. Likewise, the Pleiades begin showing up again.

In the northern hemisphere, all the stars we can see travelling over our southern skyline reach their highest point due south, just like the Sun. The Sun always peaks around noon, but the timing of each star depends on the date. There is a captivating theory that the culmination of the Pleiades at midnight was linked to the timing of Halloween. And although the particular night when it happens has changed over the centuries, this is still the time of year when the Pleiades reach their highest point more or less around midnight and are visible for much of the night.[26]

There are examples from around the world of tracking time by the Pleiades as part of everyday life.[27] Several were gathered by researchers investigating traditional knowledge in Lithuania in the last century. For instance, it was said that in November, if the Sietynas (the Pleiades cluster) is rising in the sky, it isn't yet the middle of the night. But if the Sietynas is falling, midnight has passed.[28] And in late November or early December, Russian villagers knew that when '[t]he Duck's Nest has set – it will be light soon'.[29] (Of all the names I've read for the Pleiades, this resonates most for me. The cluster does resemble a clutch of blue eggs in faint white down.)

As days pass, and the Pleiades rise, peak and set earlier, they change roles as time-givers. In Lithuania around Christmas time in December, when the Pleiades reach their highest point two or three hours before midnight, it was a sign to put children to bed.[30] Then there comes a point in late winter when they've already set by midnight.

There is a beautiful fragment from an ancient Greek poem – or possibly a complete work – which may have been composed by the poet Sappho. The poem is a lament in the voice of a woman:

The Moon has set,
And the Pleiades.
It is midnight,

STAR CLOCKS

The time is going by.
And I sleep alone.[31]

It is midnight and the Moon and Pleiades (female figures in Greek mythology) have slipped away. To me, this vivid image of time and season implies prolonged waiting and fading hope – and an empty bed around the end of winter[32] – and perhaps the longing for the warmth of a lover's body.

At first I believed the Pleiades to have been entirely erased from the sky above my home streets – that cramped and jagged space between tall buildings obscured by skyglow and crowded by the starlike light of satellites, helicopters and planes. But I am absorbing that first lesson I drew from the Irregulars. Stand still, be patient and wait for a faint blur to emerge in your peripheral vision.

After this November meeting of the Irregulars, on any clear evening I'm to be found in a safe spot at the edge of the park nearest to my home with the best local view of the southern horizon. Some nights Rosie joins me. And it's a lovely feeling to share those very rare moments when we can distinguish three or four individual stars in the Pleiades shining faintly through the veil.

On nights when the cluster is entirely hidden, we can guess where it is because we're becoming more attuned to the rusty glint of Aldebaran travelling not far behind. And as the season deepens, more brilliant followers of the Pleiades are rising up over the southern skyline earlier in the evening, giving their own sign of how far the night has progressed.

It's a chilly evening in December and I'm hunched and tired and staring at my feet. I'm scurrying home down a familiar street and happen to glance up. At the end of the road – perfectly framed between the high terraces and the red lamps from construction cranes – is the magnificent wide-shouldered figure of Orion rising over east London.

I have to miss the Irregulars this month. But I remember one of the first things Eric explained to me is that Orion returns after an absence towards the end of summer and that the constellation would be at its prime around about now. That conversation was back in October, when Orion was not yet appearing above the skyline until towards the middle of the night. Now, two months later, this huge constellation is visible for most of the night.

The stars in Orion that I can most often see from the street form the tall box shape, pictured below. The trio in the middle, which we now call Orion's Belt, was traditionally linked to the dwindling of summer and the weakening of the Sun.[33] And like the Pleiades, which rise ahead of Orion, if you know this star group's seasonal routine through the darker months, it gives another distinctive sign of the time at night.

STAR CLOCKS

In rural Russia (according to an analysis of field research mostly carried out between the 1960s and 80s), Orion's Belt (Kichigi) was among the most important groups of stars used for telling time, along with Ursa Major and the Pleiades.[34] With the Kichigi, one source said, 'You don't need clocks; we noticed the house they were above at midnight and two o'clock, and knew the time.' In late September to early October, when the Kichigi are highest before sunrise, said another, it's 'time to get up and ride to the field, to the haystack'.[35] Similarly, in Lithuania in late summer, people would get up by the sign of Orion rising; by December it was time to rise when Orion was setting.[36]

The old ways of telling time by the stars often seem to be tied to a specific landmark: a roof, a chimney, a tree; and in Iceland, the daymarks.[37] There survives a palm-sized manual for eleventh-century monks from a French monastery that explains when to sound the bells for night prayers on different dates by various star groups moving over its buildings. 'On Christmas Day,' reads one instruction, 'when you see Gemini lying almost over the dormitory, and the sign of ORION above the chapel of All Saints, prepare to stir the signal bell.'[38] Similarly, various windows in the dormitory were used as reference points. But medieval monasteries did not present a distant and unchanging skyline like the panorama of mountains around an Icelandic farm. As the historian Seb Falk points out, increasingly grand monastic buildings in this era would obstruct their residents' view of the stars. That is why the manual tells the monk that when they look for one particular set of stars 'just rising above the fifth window, near the roof' of the dormitory, they'll have to 'move back a little from the usual place towards the juniper bush'.[39]

Nearly a month after the glorious sight of Orion between the construction cranes stopped me in my tracks, this huge figure is higher in the sky earlier in the evening. And that means we're more aware of Sirius the Dog Star, which chases after Orion. Sirius is the brightest star in the firmament, but in London skims low to the rooftops. It twinkles so energetically that at a glance you could mistake it for one of the helicopters hovering over the city. This brilliant object is my next focus because, as we turn into the new year, Sirius reaches its highest point around about midnight.

The canal on this January evening is a thin rectangle of black glass flecked with a few papery yellow leaves. Over the bridge and under the bare trees, the air is dark and dank. It's still early but feels like deep night. Halfway to the Hub, astoundingly, we hear the breathy hoot of a tawny owl.

The Hub is fairly empty tonight. Yet again this dismal winter, there seems to be little prospect of viewing stars. But a few of us Irregulars buy a drink at the counter beside half-unpacked telescopes.

Over a hot tea, I ponder how remote seem the 'dog days' of late summer – the hottest time of year, which long ago was linked to the appearance of the Dog Star.[40] For the ancient Egyptians, this most brilliant of stars was the goddess Sopdet. Her return to the sky was associated with the annual flooding of the Nile, which enriches the land. Lately I've been discovering the story of Sopdet and the role she plays in how the modern hours came into being.[41] That is, why our dominant system divides the day-night cycle into two sets of twelve hours and not twenty or thirty or any other number of parts.

More than four millennia ago, Sopdet was primary among the cycle of special stars by which the ancient Egyptians kept track of the weeks – which lasted ten days – in the civil calendar. Each of these bright stars disappeared for around seventy days. Each ten-day week began when one returned to the sky at sunrise. Every morning that star would appear a little earlier until it was replaced by a new rising star, heralding a new week, and then came the next star and the next. At an uncertain point in history, these special weekly stars were understood to divide the darkness into twelve phases called *wnwt* (pronounced, more or less, 'wenut'). This appears to be the first inkling we get of the Egyptian hours which would eventually influence the number of our own hours.

We know about the *wnwt* from the star timetables that have survived, for example, on temple ceilings. But we don't know their purpose. Perhaps priests watched the *wnwt* stars to keep track of the journey of the sun god Re through the twelve chambers of the Underworld, the Duat, as he fights off demons every night. Maybe the priests were timing temple rites and prayers to help Re pass safely through each gate, and to know when to prepare to celebrate his rebirth at dawn.

The star tables are tools for the dead as well as the living, proposes the leading scholar Sarah Symons.[42] Not only do the tables appear on temple ceilings but on tomb ceilings and inside coffin lids. These maps of the motion of celestial bodies may have been intended as a guide for the deceased on how to behave when they were reborn as a star. And they weren't only included in human tombs. The one surviving star map that shows how to tell time by Ursa Major – Meskhetiu, the Foreleg – is carved inside the granite sarcophagus of a sacred bull.[43]

Some thirty centuries or more before our present, the Egyptians started dividing the daylight, too, into twelve *wnwt* measured by sun shadows. Most historians now agree that the ancient Egyptian practice of counting twelve night-*wnwt* and twelve day-*wnwt* appears to be at the root of why we came to count two sets of twelve hours.[44] (Europe would inherit a version of this division from the Greeks through the Roman Empire and Byzantine world.) By extension, it is often said that the ancient Egyptian hour, *wnwt*, is the origin of our modern twenty-four-hour day. But as logical as it sounds, is this too big a leap?

Sarah points out that the ancient Egyptians would not have thought of time as we do, that is, as an abstract and continuous stream of empty identical units. Ultimately each *wnwt* was defined by a concrete event: a special star rising from the local horizon, say, or a shadow touching a mark in the stone.[45] What's more, the night-*wnwt* and day-*wnwt* did not make a seamless cycle of twenty-four uniform hours. Because, like the medieval European hours that inherited their basic pattern, the *wnwt* stretched and shrank in span between dawn and dusk as the balance of darkness and daylight tipped back and forth through the year.[46]

The role of Sopdet in the history of the hours is a reminder that the twenty-four standard hours that define time around the globe now are, of course, not historically or culturally universal. The ancient Egyptian *wnwt* help explain why we divide the day into two sets of twelve hours. Yet, myriad practices from multiple cultures over thousands of years stand between the *wnwt* and our modern scientific hours.

Looking out at glittering nocturnal London and waiting for Sirius/Sopdet to appear, it seems obvious that in many ways that stark contrast between day and night has dissolved for us, both in the brightly lit material world and the permanent daytime of the internet. Naturally, it's harder to feel so keenly that older distinction between day and night as separate states. The smooth circling of the clock hand over the abstract, identical hours represents, so to speak, how the rhythm of the day has lost its texture for us.

A few months later, when I speak with Sarah Symons about Sopdet and the hours, she draws out another detail that exposes just how differently I think about the measure of time as a modern secular European.

'If time at night is told by stars, and time during the day by shadows,' she says, 'then there are gaps in between.' These are the interludes of soft light and fiery skies just after sunset and before sunrise when there are neither direct sunbeams nor time-giving stars.[47] In Egypt this brightest phase of twilight unfolds much faster than in southern England. Nevertheless, Sarah proposes, 'it would have been a time without time'. For the ancient Egyptians, the day and night seem to have been divided by a timeless margin.

Back in London at this January meeting of the Irregulars, the dark, shadowless day has dissolved seamlessly into a starless gloom. My hope had been to see how Sirius flickers with colour when magnified through a telescope. But the grim cold, the spongey cloud, the potential for drizzle cast a gloomy prospect and I head home early.

As my train rattles through a dark tunnel deep under the city, I have the sense that journeying to the monthly meetings is its own kind of ritual. Whether or not the stars show, I'm committing to honouring the night as its own special state. And maybe next time the weather will be better.

A few nights later, Rosie and I are walking home through our neighbourhood in the shivering cold. The tarmac is starting to glisten with frost, and at last the sky is clear. Our main street, though, is peppered with so many flaring lights, it's like walking through a miniature star cluster. We haven't a hope of spotting the Pleiades, nor even the Plough. But when we turn to cross the street and glance south to check the traffic, we see the sparkling light of Sirius about to cross in the other direction. It'll be a while yet before midnight and already this brightest star in the firmament is nearly over Noon Street.[48] The stars and the season are moving on.

On the way to the Hub for the February meeting, a blackbird is singing fitfully in the dusk shadows by the canal. A tiny buff-coloured moth flutters into the lamplight. Soft rain patters and drips from the tips of the buds on the branches in the avenue of trees. The year is waking up and it won't be long now until the evenings are light again.

Inside the Hub, I find the Irregulars have given up on the weather and are dreaming about what would make an ideal public observatory in London. Well, there's the top of the BT Tower, Eric says. That might help you see a little better above the tall buildings and streetlamps but wouldn't make a difference to the skyglow. Seriously, though, what if you could place public telescopes all around the city?

In my own version of this fantasy, the metropolis would have maps everywhere showing how to tell time by the stars. We would paint charts onto train-station walls and weld the stars of the Plough spinning around the North Star into railings on the street. If the city were full of star clocks, how differently would we sense the hours? How much more would we open our mind and senses to the nocturnal sky-world? What care would we take to reduce the baffle of skyglow overhead?

It's been five months now since I first joined the Irregulars, and I reflect on how much I've gained and learned. True, I'm no Gabriel Oak. I remain a novice time-reckoner by the stars

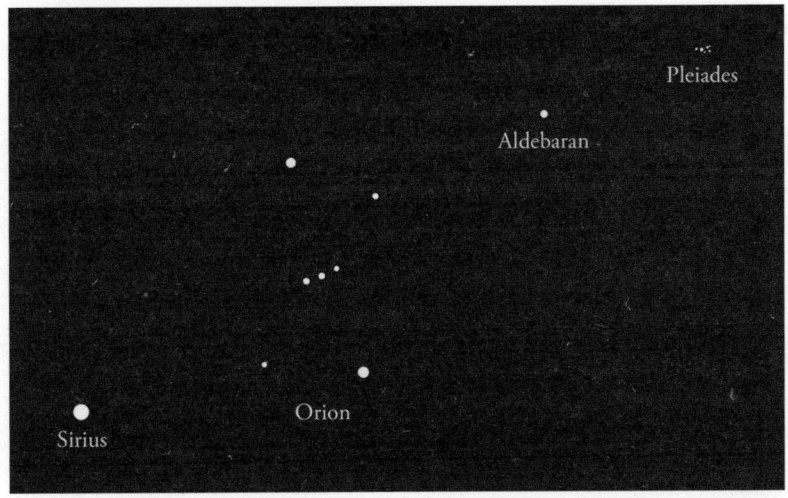

and am still unacquainted with very large regions of the night sky. But by investigating what one bright object is doing each month, I've come to recognise this one piece in the whole pattern Gabriel saw from the hill: 'The Dog-star [Sirius] and Aldebaran, pointing to the restless Pleiades, were half-way up the Southern sky, and between them hung Orion.'

Other than Orion, those points of light were anonymous to me back in October. Now I can identify them as a train of bright bodies, each of whom has their own season and hour of triumph. Even in the city, I can often glimpse their basic pattern, pictured above, moving over the southern horizon in a grand procession.

One of the astronomer's tricks I've learned from the Irregulars is to take a quick and literal rule of thumb by measuring objects with my hand and fingers. For the average adult, if you make a fist and hold it at arm's length, with your index and little fingers stretched apart, the distance between the two fingertips is around fifteen degrees. That's how far a star appears to travel over the course of roughly a clock hour. And it's a pleasing habit to map with my hand how much Orion has moved against the side of our block of flats or over the chimney pots while I've been out for the night.

To gesture to the sky – to orient yourself in time and space by mapping the movement of the stars against the local skyline with your own lively digits – feels remarkably different from casting a glance at a machine. It's among those other ways of gauging time by the body and the senses – by the shrinking of a shadow, the glow over a daymark, the changing colours of the sky and air – that binds us to our present world.

Tracking the motion of the stars by the body may be a very old practice indeed, as I will later learn from Sarah. In the tombs of the ancient pharaohs, the star tables are accompanied by a picture of a seated human. The image has no instructions and its interpretation is debated. But it clearly maps the changing positions of the stars against the figure's eyes, ears, shoulders and heart. Not that the ancients saw exactly the same night sky that we do, because the sky dome appears to shift and change very gradually through the ages. That means the stars we call the Plough didn't wheel around the 'North Star' three thousand years ago. Instead there was a starless gap over the celestial pole. But we do see many of the same stars and star groups. To the ancient Egyptians, Sirius was Sopdet, as we've learned, and the Plough was Meshketiu.

The more time I devote to the starry sky, the more palpably I'm aware that we share this glittering commonwealth with every earthling in our present and every earthling in the past. But where just a few decades ago Richard Proctor could see hundreds, if not thousands of stars, we are left with only the brightest.

STAR CLOCKS

Over this long gloomy winter I've become more and more enchanted by what remains of London's night sky. I've made it a ritual after supper to see if the Great Bear has appeared over the rooftops, like a sunk boat slowly rising from the sea. And I've begun to feel the slow-wound suspense as the train of stars led by the Pleiades rises earlier and earlier in the evening. When I first joined the Irregulars at the October meeting, I barely knew about the Pleiades cluster, let alone imagined I could spot it in London. But from them I've learned to exploit averted vision and, even from the carpark, can sometimes spy a soft blur where it should be. I feel as if I've slightly shifted state, becoming more alert to subtle flickers and more able to still my senses.

A few evenings ago, I was standing motionless at the edge of Regent's Park, gazing upward, when a bushy-tailed fox came very close, as if joining me to take in the world. The fox stood with me for half a minute, studying something my eyes could not see, then bounded away into the darkness without a sound.

Epilogue

Long ago I began collecting stories about how people marked times of day without complex tools. And in tandem I started devoting attention to the rhythms of life I'd lost touch with in my urban habitat: the dawn birdsong, the opening of day's-eyes, the lengthening of shadows, the golden glow in the sky, the silent swoop of the bat. To my delight and surprise, that lovely pursuit has slowly burgeoned into a way of life. Wherever I go now, I'm less distracted from the present moment by my phone screen and much more alert to tiny details in my surroundings. I'm growing more adept at distinguishing the dandelions from the almost-dandelions: the hawkbits and cat's ears and other lookalikes[1] that unfold or fold up at different times. In winter I'm likely to press my ear to a suburban hedge to listen for sheltering sparrows. I'm tuned into the tints of green in twilight skies, and alive to the opulence of the stars. As I go about the world, I have a stronger sense of being enveloped by a great weave of rhythms in the more-than-human world, of living inside the sundial and under the revolving dome of the night sky.

THE FULLNESS OF TIME

All throughout this adventure, Rosie has been with me and pointing out joyful things I haven't noticed: the Moon glowing through the clouds, the 'chalky light' of a summer solstice dawn, the detail in the robin's song. We've both become more attuned to how sunbeams and stars move over the rooftops through the day and year.

Mum and Dad have always been a part of this too. When, to my delight, 'the butterfly flutters by', I'm hearing Mum's voice. When I look for a flower folding up, I find my stance is like Dad's when he carefully studied the fields.

Where I grew up, there's a red-brick coach arch in the corner of the farmyard. Some years back, my parents hung up pieces of old kit on the walls underneath the arch: two great rusting scythes, loops of chainwork, horseshoes big and small. If you stand in the farmyard looking out on a fine evening, the arch frames the sunset sky, and the golden light throws long and intricate shadows of the old tools onto the bricks. This is a very familiar sight, but only in mature adulthood did it really occur to me that it's seasonal: sunset is directly in line with the arch only around the middle of summer. When I mentioned that to Dad, if he wondered why I'd cottoned on so late in life, he didn't let me know. Instead, he showed me the different places where the Sun sinks behind the hedgerows through the year. Dad had lived his whole life out in the fields, and this was utterly obvious to him. But it was a revelation to me.

There is so much we don't ask, and so much out there to discover.

Acknowledgements

The person I have learned from, talked to, and shared most of these adventures with, of course, is Rosie. She has encouraged me all the way with splendid ideas, conversations and adventures – and her incredible practical ability to do things like drive us safely through an Icelandic snowstorm. She has helped sustain this project with unbelievable endurance, exuberance, curiosity, honesty, laughter and love.

As I said at the beginning, this book has increasingly become about remembering my mum and dad, Pat and Stephen. It's my roundabout way of honouring and making the most of the gifts they gave us, particularly their delight in telling stories and their appreciation of details in the world.

My parents' legacy continues in many other ways under the love, care and leadership of my sister Lizzie with her partner Kevin and children Ophelia and Joseph. Lizzie is out in the fields, testing grain with her teeth or listening for the skylark. She has the knowledge that Dad and Mum had – and continues to teach me.

I am blessed, too, to have the support of my magnificent mother-in-law Dawn and Rosie's family. My life has been buoyed by their love, warmth and teasing through the last quarter-century.

This has been a time of overwhelming sadness for us all. Not long after we lost Mum and Dad, we lost my beloved aunts Susie and Pam. Rosie lost her lovely uncle, James, and step-grandmother, Lia. And she lost her father, Stephen, who was such an important person in our lives. I am so grateful for their love and support, and all they taught us.

Now to thank my excellent friends for their inspiration, fun and encouragement along the way, especially Jmeel, Shab, Sabih, Fraser, Sharmini, Jan, Robbie, Mike, Kenan, Sally, Matt, Jinny, Rebecca, Rachel A, Rachel P, Dan, Rhys, Tass, Ines, Anneli, Julie, Leon and Rod.

Then there are the people who have worked with me directly on *The Fullness of Time*. I am deeply indebted to writing coach Kathy Gale of Gale & Co. for her great insight and consistent belief in me. Her galvanising, challenging, steadying mentorship has been profoundly transformative.

I'm hugely thankful for the piloting, patience and brio of my agents, Rowan Lawton and Farley Chase, and their teams, especially Eleanor Lawlor. And for the great wisdom, support and invigorating ideas given by my editors, Jasmine Horsey at Bloomsbury and Courtney Young at Riverhead. I've learned enormously from them. It was Jasmine who gave me the pivotal advice to seek delight and convey the sensory experience.

I've very much valued the brilliant clarity and thoughtfulness of my copy-editor, Kate Quarry. I am indebted to fact checker Lucie Kroening for her rigour and insight. And to the Scottish Gaelic expert Mìcheal Bauer for his careful eye. I am hugely thankful to the designer and illustrator Charlotte Phillips for her splendid cover artwork, the astronomer and

ACKNOWLEDGEMENTS

illustrator John A. Paice for so thoughtfully redesigning and refining the star maps, and the illustrator Carmen Balit for perfecting the diagrams. I've greatly appreciated the brilliant support, too, of Francisco Vilhena, Gurdip Ahluwalia and the wider publishing team.

The Fullness of Time would truly not have been possible without the generosity of many experts. I am especially beholden to Mario Arnaldi for sharing such a treasury of knowledge, and for his enthusiasm. I have benefited vastly, too, from the expertise of Gísli Sigurðsson, Birna Lárusdóttir and Stephen Corfidi, who have given amazingly rich and patient insights in response to my relentless questions.

My wholehearted thanks to those experts who have made really significant suggestions and contributions, given in-depth interviews or commented on longer draft sections of the book at various stages of development: Anne Lawrence-Mathers, Thomas Hockey, Sara Schechner, Karlheinz Schaldach, Ross Collery, Sylvain Aubry, Alex Webb, Beverley Glover, Staffan Müller-Wille, Frances Dunlop, Fiona Mackenzie, Ben Jones, Josh Pollard, Stuart Mosscrop, Steve Bell, Imad Ahmed, Sarah Symons, Dominic Couzens, Fabio Silva, Uta Wolfe, Bennett Schwartz, John A. Paice, Ian Stone, Zoe Hutin, Eric Emms, Nick Joannou, and Marian, David, Clare, Danny, Simon and Archie at Court Lodge Farm. I'm particularly grateful to have had the opportunity to correspond with the late Þorsteinn Vilhjálmsson.

Special thanks for their expert insights to Lilla Kopár, Fergus Kelly, Jole Shackelford, Taha Yasin Arslan, Stacey Harmer, Mike Meynell, Norman Solomon, Catherine Hohenkerk, Mike Frost, Ian Barthorpe, Douglas Parker, Peter Beckenham, Richard North, Sarah Lyons Chase,

Sacha Stern, Fiona Rosher, Árni Björnsson, Gareth Jones, Emma Stone, Halla Steinólfsdóttir, Jesper Kårehed, Mark Waddell, Nicola Loaring, Sigrún Kristjánsdóttir, Frank King, Ken Mondschein, Simon Saville, Maggie Couvillon, Gert Wilkens, Pru Manning, Isabelle Charmantier, Kevin Walker, Alex Prendergast, and Jem Finer, Phil Serfaty and the Longplayer Trust team.

Many thanks, too, to all those who've answered my queries, corrected my misunderstandings and shared references, especially Sian Prosser, Helen Klus, Helga Vollertsen, Ágústa Kristófersdóttir, Aðalsteinn Hákonarson, Torfi Stefán Jónsson, Gunnlaugur Björnsson, Ragnheiður Mósesdóttir, Rod Dimbleby, Elin Pihl, Anne Courtin, Dagrún Ósk Jónsdóttir, Taj Bhutta, Johan Anton Wikander, Kelly Fitzgerald, Karen O'Toole, Theresa Kelly, Janken Myrdal, Jayne Carroll, Gordon Love, David Jacobs, Claire Ryder, Sue Manston, Katherine Wodehouse, Kevin Murray, Geoffrey Keating, Kevin J. McGowan, Kathi Borgmann, Andrew Foley, Till Roenneberg, Sarah Lo, Dave Clark, Louise Lawton, Brigitte Steger, Kent Wildlife Trust, the teams at Cambridge University Botanic Garden and Walworth Garden, the very helpful librarians at various institutions, and many, many others along the way.

What exactly sparked the enthusiasms behind this book is not obvious to me. But I remember one important conversation about traditional ways of dividing time and space among pastoral societies with the anthropologist Malcolm Ruel, my undergraduate supervisor at Clare College, Cambridge. He encouraged me to notice and value the everyday traditions I grew up with, and use my visual skills, rather than just ponder scholarly texts in the abstract. That may be the origin of my decades-long research and creative projects investigating

ACKNOWLEDGEMENTS

aspects of how we experience, think about, map and measure time.

I'm grateful to Alfredo Cramerotti for commissioning my first solo projects on time themes, Roman Krznaric for the confidence he showed in these early endeavours, and Alex Bowler for long ago suggesting I write about time and giving me invaluable feedback. It was a pleasure to collaborate with the team during my curatorial residency as 'timekeeper' at the Petrie Museum of Egyptology – especially with the marvellous Helen Pike, who is very sadly missed. It was a privilege and a pleasure, too, to work with Laura Wilson, Polly Staple and Razia Begum at Chisenhale Gallery on *Stereochron Island*, my project as artist-in-residence in Victoria Park, London.

If I've missed anyone in this effort to list all the fine people who have helped and contributed, my heartfelt apologies to you.

I'm not the first to explore these themes for a broader readership in recent years. Like many others, I travel in the wake of Jay Griffiths' magnificent *Pip Pip: A Sideways Look at Time* (1999).[1] Tristan Gooley is perhaps prime among contemporary authors of powerful poetic and practical guides to connecting with the natural world by observing its rhythms through the day and year.

While my book is not an academic work, I have tried to learn from the recent move in scholarship to investigate the diversity of time-telling practices in everyday life. The shift is inspired in part by Paul Glennie and Nigel Thrift's study of timekeeping in England and Wales from 1300 to 1800, *Shaping the Day* (2009),[2] which reveals a complicated, fluctuating and varied picture on the ground. Another magnetic influence today is Henri Lefebvre's concept of 'rhythmanalysis', which invites sensory research into the many different rhythms of urban life, including those of birds, flowers and shadows.

Among the innumerable influences I've absorbed over the years, one resource I've turned to repeatedly is Nina Gockerell's 'Telling Time Without a Clock' (1980). Histories of Western timekeeping and time measurement typically prioritise innovations in precision instruments. In contrast, Gockerell's strangely overlooked but gloriously full essay gathers everyday methods of telling time without complex tools among laypeople, particularly in rural Europe.

It has been a complete joy to roam far beyond my own field and learn from so many different experts and wide-ranging sources. For me, and I hope for you, the unusually broad range of chapter themes in this book is one of its pleasures. I've aimed to take complicated topics and introduce them simply for other lay enthusiasts like me, and to go beyond the abstract through experience. Along the way, I am bound to have made slips or missed significant details. I've done my very best to pin down the facts and, wherever possible, made every effort to check my account with experts. Any errors or misunderstandings are, of course, entirely my own. I've spared you much longer stories about convoluted research processes, and in places have simplified, reordered or merged events, recollections or conversations for clarity. Some ideas have played in my mind for many years, and I hope I haven't unwittingly repeated others' words without acknowledgement.

The aspects of everyday life that seem obvious and natural to us may not be to future generations. Just so, it may not have occurred to people in the past to record their routine

ACKNOWLEDGEMENTS

ways of telling time, even if they had the means. Investigating what people actually did and thought has been one of the most absorbing and challenging aspects of this project. As one historian exclaimed in our conversation about medieval rural rhythms of the day, 'So much uncertainty!' Inevitably, not all specialists will agree with speculations I've made or relayed, and I welcome the discussion.

Excitingly, this is a tiny part of a much bigger picture, and I hope to hear from readers with insights to share.

A note on the text

I've used words like 'time', 'timing' and 'telling time' broadly, rather than attempting to strictly define such slippery concepts that are dependent on context, culture and historical moment.

I've noted the scientific name for a particular species of plant or animal only where I think it could be confused with another. Except, that is, for chapter two, where I've given scientific names for all the plants with rapid movements.

In drawing on histories and historical texts, I am of course making use of their evidence without endorsing any politically extreme or prejudiced views that may belong to their authors.

When quoting from earlier forms of English, my guiding (if subjective) principle has been ease of reading. If a quotation seems accessible, particularly when read out loud, I've tended to give the original. When I've given a modern translation instead, I've aimed to provide a reference in the endnotes so you can seek out the source.

When I've quoted or cited a work that has multiple editions with different chapter and page numbers, I've not always given a detailed reference in the endnotes. This is especially the case with Hardy's novels, which are freely accessible in online editions that make it easy to search within the text for a word or phrase.

Notes

INTRODUCTION

1 I am grateful to the meteorologist Douglas Parker for his insights on these examples.
2 For this complex history see, for example, Vanessa Ogle, *The Global Transformation of Time, 1870–1950*, Harvard University Press, 2015, and David Rooney, *About Time: A History of Civilization in Twelve Clocks*, Viking, 2021.
3 'On a Sun-Dial', 1827. Sadly, Hazlitt's wider argument reveals his social prejudices.
4 Although I've not made a thorough survey, I'm struck by the abundance in Hardy's novels of signs of the time of day and night from the land and sky and fellow creatures. In *The Mayor of Casterbridge*, for example, Abel Whittle reckoned time 'by the sun; for I've no watch to my name'. In *The Woodlanders*, Grace realised three hours or more had passed by the sunbeams reaching fully over the treetops. In the chapters that follow, we'll return to other examples from Hardy that I've listed in the Introduction.
5 Ulf Büntgen et al., 'Plants in the UK flower a month earlier under recent warming', 2022, *Proceedings of the Royal Society B*, 2 February 2022, http://doi.org/10.1098/rspb.2021.2456; Chris Wyver and Laura Reeves, 'Plants are flowering a month earlier – here's what it could mean for pollinating insects', *The Conversation*, 4 February 2022, https://theconversation

.com/plants-are-flowering-a-month-earlier-heres-what-it-could-mean-for-pollinating-insects-176324

1. DAWN SONG AND BAT FLIT

1 The word *wōma* means 'noise, howling, tumult' (John R. Clark Hall, *A Concise Anglo-Saxon Dictionary*, Cambridge University Press, 1916, p. 360). That would associate *dæg-wōma* with the cacophony of sounds at daybreak, although more often its poetic associations seem to be with the morning light, as in line 344 in the tenth-century poem *Exodus*. I'm grateful to the scholar Lilla Kopár for sharing these details.
2 Some early singers may have fallen quiet by now, but I was not aware of that amid the intense chorus.
3 I've found repetitions of the report in various British and North American publications. For example, *The Athenæum ... for the Year 1857*, No. 1559, London, 12 September 1857, p. 1145; and *The Family Friend*, Ward and Lock, London, Midsummer 1861, p. 348.
4 *Graham's Illustrated Magazine*, Vol. 53, No. 1, Philadelphia, July 1858, p. 278.
5 See Dominic Couzens, 'Designed by nature: The secrets of the dawn chorus', *The RSPB Magazine*, Spring/Summer 2025, p. 11.
6 See ibid. Also Dominic Couzens, *Songs of Love & War: The Dark Heart of Bird Behaviour*, Bloomsbury, 2017, pp. 21–8.
7 See John Brand and Henry Ellis, *Observations on Popular Antiquities*, Chatto and Windus, London, 1877, pp. 323–4; Ebenezer Cobham Brewer, *Wordsworth Dictionary of Phrase and Fable*, Wordsworth Editions, 2001, p. 263; and Kevin Birth, 'The Regular Sound of the Cock: Context-Dependent Time Reckoning in the Middle Ages 1', *KronoScope*, 11(1–2), 2011, https://doi.org/10.1163/156852411X595305, especially pp. 127 and 131.
8 *Cinnlae Amhlaoibh Uí Shúileabháin: The Diary of Humphrey O'Sullivan*, Part 2, Michael McGrath (ed.), Irish Texts Society, London, 1936, pp. 354–7.
9 Henry Bourne, *Antiquitates Vulgares; or, the Antiquities of the Common People*, J. White, Newcastle, 1725, pp. 37–8. Also see Brand and Ellis, *Popular Antiquities*, p. 322.

NOTES

10 Jacqueline Simpson and Steve Roud, *A Dictionary of English Folklore*, Oxford University Press, 2000, p. 74.
11 'The Morning Quatrains', in *The Poetry of Charles Cotton*, Vol. 1, Paul Hartle (ed.), Oxford University Press, 2017, pp. 293–6.
    The rooster and the lark are closely linked in Cotton's poem. And I notice that while the lark is the 'herald of the morn' in *Romeo & Juliet*, the *cock* 'is the trumpet to the morn' in Shakespeare's *Hamlet*.
12 For example, for what appears to be an early Irish reference to morning and evening milking, see Fergus Kelly, *Early Irish Farming: A Study Based Mainly on the Law-Texts of the 7th And 8th Centuries AD*, Dublin Institute for Advanced Studies, 1997, pp. 38–9.
    Traditionally, milking may have been reduced or stopped in winter (see ibid., pp. 41 and 65). For a portrait of the rhythm of the day and year on a nineteenth-century dairy farm in southern England, see Hardy's novel *Tess of the D'Urbervilles* (1891).
13 § (Section) 8 in Thomas Charles-Edwards and Fergus Kelly (eds), *Bechbretha: An Old Irish Law-Tract on Bee-Keeping*, Dublin Institute for Advanced Studies, 1983, pp. 54–5.
14 Fergus Kelly generously shared this with me. For the likely timing, he made a comparison with the tenth-century glossator Cormac Ua Cuillennáin (Cormac's Glossary § 554).
15 'Evening Quatrains' in *Cotton*, Hartle, pp. 300–1.
16 The time when chickens are cooped is one of two potential meanings of the word. See 'Cockshut, *N.*, Etymology', *Oxford English Dictionary (OED)*, Oxford University Press, June 2025, https://doi.org/10.1093/OED/1067706660
17 John Florio, *A Worlde of Wordes, Or Most Copious, and exact Dictionarie in Italian and English*, Arnold Hatfield, London, 1598, p. 56.
18 See Laura K. Lawless, 'Entre chien et loup', Lawless French, undated, https://www.lawlessfrench.com/expressions/entre-chien-et-loup
19 Nina Gockerell, 'Telling Time without a Clock', in Klaus Maurice and Otto Mayr (eds), *The Clockwork Universe: German Clocks and Automata, 1550–1650*, Smithsonian Institute and

Neale Watson Academic Publications, Washington D.C. and New York, 1980, p. 138.
20 Ulrich Bräker, *The Life Story and Real Adventures of the Poor Man of Toggenburg*, Derek Bowman (trans.), Edinburgh University Press, 1970, p. 67.
21 Couzens, 'Designed by nature', p. 11.
22 Couzens, *Songs*, p. 31.
23 'Night Flight: Tracking Data Reveal When and Why Songbirds Begin Their Massive Journeys and How They Decide to Leave', Smithsonian, 1 May 2023, https://www.si.edu/newsdesk/releases/night-flight-tracking-data-reveal-when-and-why-songbirds-begin-their-massive; Fiona McMillan, 'From dung beetles to seals, these animals navigate by the stars', *National Geographic*, 4 November 2019, https://www.nationalgeographic.com/animals/article/stars-milky-way-navigation-dung-beetles
24 Cotton refers to the owl and nightingale as *'Nyctimine'* and *'Philomel'*.
25 In early 2025 (after this trip), plans were announced to expand Heathrow Airport to take 66 million more passengers a year. The prospect is alarming, not least in terms of the potential increase in carbon emissions, noise and air pollution.
26 For an ongoing study of trends in UK insect populations, see the nationwide Bugs Matter citizen science survey, led by Kent Wildlife Trust and BugLife, at buglife.org.uk. For trends in bat populations, see the National Bat Monitoring Programme Reports provided by the Bat Conservation Trust, bats.org.uk
27 Alyson Brokaw, 'Blind as a bat? No such thing', Bat Conservation International, 14 June 2024, https://www.batcon.org/blind-as-a-bat-no-such-thing
28 See, for example, Diego Gil et al., 'Birds living near airports advance their dawn chorus and reduce overlap with aircraft noise', *Behavioral Ecology*, Vol. 26, No. 2, March–April 2015, https://doi.org/10.1093/beheco/aru207; and Dustin G. Reichard et al., 'Urban birdsongs: higher minimum song frequency of an urban colonist persists in a common garden experiment', *Animal Behaviour*, Vol. 170, 2020, https://doi.org/10.1016/j.anbehav.2020.10.007

NOTES

29 See, for example, Domhnall Finch, Henry Schofield and Fiona Mathews, 'Traffic noise playback reduces the activity and feeding behaviour of free-living bats', *Environmental Pollution*, Vol. 263, Part B, 2020, https://doi.org/10.1016/j.envpol.2020.114405

30 See 'Guidance Note GN08/23: Bats and Artificial Lighting at Night', Institute of Lighting Professionals (ILP) and Bat Conservation Trust, ILP, Rugby, 2023, https://theilp.org.uk/resource/gn08-bats-and-artificial-lighting-pdf.html. For research on the impact of artificial light on the commuting behaviour of lesser horseshoe bats (now mostly found in Wales and Southwest England), see Emma Stone, Gareth Jones and Stephen Harris, 'Street-lighting disturbs commuting bats', *Current Biology*, Vol. 19, 2009, https://doi.org/10.1016/j.cub.2009.05.058, pp. 1223–7. My thanks to Elliot Newton, Emma Stone and Gareth Jones for their guidance on this topic. Any misunderstandings are my own.

31 Although the zoologist Emma Stone told me that, in her experience, the tower and nave of a church are far more likely than the belfry.

32 Todd M. Jones et al., 'Phenotypic signatures of urbanization? Resident, but not migratory, songbird eye size varies with urban-associated light pollution levels', *Global Change Biology*, 29(23), 2023, https://doi.org/10.1111/gcb.16935

33 See the Bird Collision Prevention Alliance website, stopbirdcollisions.org

2. DAY'S-EYES AND TURNSOLES

1 'Daisy, *N.*, Etymology', *OED*, June 2025, https://doi.org/10.1093/OED/2125426858

2 Lines 182 to 184 in the Prologue to *The Legend of Good Women*, A.S. Kline (trans.), 2008, Poetry in Translation, https://www.poetryintranslation.com/PITBR/English/Good Women.php

3 The daisy (*Bellis perennis*), common marigold (*Calendula officinalis*), chicory (*Cichorium intybus*), dandelion (*Taraxacum* species) and sunflower (*Helianthus annuus*) are all in the family Asteraceae. Their flower-heads are actually tight clusters of

multiple flowers. Their 'petals' are flowers, too, known as ray florets. For brevity, I sometimes refer to their flower-heads as flowers.

4 There are hundreds of micro-species of dandelion. I haven't been able to identify this particular dandelion, which I photographed in the English Midlands.

5 For local names, including one o'clock and shepherd's clock, see Geoffrey Grigson, *The Englishman's Flora*, Reader's Union, London, 1958, p. 392.

6 The name is understood to derive from the children's game of pretending to tell the hour by the number of puffs it takes to blow the seed away. For the plant's variety of local names, see Grigson, *Flora*, pp. 393–4; also T. F. Thistleton-Dyer, *The Folklore of Plants*, Chatto & Windus, London, 1889, p. 123.

7 Thistleton-Dyer, *Folk-lore*, p. 124.

8 Another example is *Ornithogalum umbellatum*, whose local names include eleven o'clock lady and twelve o'clocks (Grigson, *Flora*, pp. 404–5).

9 Pliny the Elder, *Natural History*, 18.252; also see 2.109 and 22.57.

10 Thomas Hill, *Profitable Art of Gardening*, Henry Bynneman, London, 1579, pp. 93–4.

11 John M. Wilson (ed.), *The Rural Cyclopedia Q–Z*, Vol. 4, A. Fullarton and Co., Edinburgh and London, 1849, p. 639.

12 John Gerard, *The Herball Or Generall Historie of Plantes*, John Norton, London, 1597, pp. 612–4. This is the source of all quotations from Gerard in this chapter. (For ease of reading, I have swapped the letters 'u' and 'v' in his spellings. For example, I've changed 'vnshorne veluet' to 'unshorne velvet'.)

13 'Heliotrope, *N*., Sense 1.a', *OED*, June 2025, https://doi.org/10.1093/OED/3459439091; 'Turnsole, *N*., Sense 2.c', *OED*, December 2024, https://doi.org/10.1093/OED/1037875179; and Henry N. Ellacombe, *The Plant-Lore and Garden-Craft of Shakespeare*, Edward Arnold, London and New York, 1896, pp. 164–7.

The common marigold opens in the morning and closes at the end of the day with regularity. It was also believed to *turn* with the Sun – which it does not do, as far as I've been able to discover. See Thistleton-Dyer, *Folk-lore*, p. 123.

NOTES

14 For the sunflower's symbolism, and in comparison with the marigold as an emblem, see Sam Segal and Klara Alen, *Dutch and Flemish Flower Pieces*, Vol. 1, Brill, Hes & De Graaf, Leiden and Boston, 2020, pp. 54–5 and 60–4; and Elizabeth Nogrady, 'Young woman holding a sunflower', 2024, *The Leiden Collection Catalogue*, 4th ed., Arthur K. Wheelock Jr. et al. (eds), New York, 2023–, https://www.theleidencollection.com/archives/artwork/BH-100_young-woman-holding-a-sunflower_2024.pdf

15 Joscelyn Godwin, *Athanasius Kircher's Theatre of the World: His Life, His Work, and the Search for Universal Knowledge*, Thames & Hudson, London, 2009, pp. 18 and 21.

16 See Mark A. Waddell, 'Magic and artifice in the collection of Athanasius Kircher', *Endeavour*, Vol. 34, No. 1, 2010, pp. 30–4; Paula Findlen (ed.), *Athanasius Kircher: The Last Man Who Knew Everything*, Routledge, 2004; and Godwin, *Kircher's Theatre*.

17 *Magnes sive de Arte Magnetica* (*The Lodestone, or the Magnetic Art*), first published in 1641.

18 For a compelling analysis of Kircher's wider intentions, see Waddell, 'Magic'.

19 Solar-tracking movement in plants is called heliotropism. In the sunflower, the heliotropic motion of the bud through the day is driven by the stem growing asymmetrically: the east side expands more than the west. The chronobiologist Stacey Harmer explained to me that, 'As the plant reaches reproductive maturity, the stem becomes woody and lignified and is unable to grow. At this point the head becomes stuck pointing in one direction. In our experiments, we found this direction is determined by where the Sun rises.'

I asked Stacey why several of the fully grown sunflowers in my neighbourhood are ignoring the morning Sun. Some varieties of sunflower (*Helianthus annuus*) are more variable in the direction they end up pointing, she replied. There may be other reasons too. One significant influence, she suspected, is that if a sunflower is grown in a spot where the skyline is obscured by tall buildings, that may have the effect of inhibiting its solar-tracking movement and, in turn, weakening its eastward orientation at maturity.

20 'The Garden', published in 1681, after Marvell's death in 1678.
21 The timings follow those given in B. G. Gardiner, 'Linnaeus' Floral Clock', *The Linnean*, Vol. 3, No. 1, Linnean Society, London, 1987, p. 28.
22 Stephen Freer (trans.), *Linnaeus' Philosophia Botanica*, Oxford University Press, 2003, p. 297.
23 Freer, *Philosophia*, pp. 293–6.
 Linnaeus's floral clock table is given online at 'Horologium Florae according to Philosophia Botanica, 1751', The Linnaeus Garden, Uppsala Universitet, https://www.uu.se/en/linnaeus-garden/our-plants-and-attractions/our-plants/linnaeuss-floral-clock/philosophica-botanica
24 Anton Kerner von Marilaun, *The Natural History of Plants: Their Forms, Growth, Reproduction, and Distribution*, Vol. 2, F.W. Oliver (trans.), Blackie & Son, London, 1895, p. 215.
25 'Linnaeus's floral clock – Horologium', The Linnaeus Garden, Uppsala Universitet, https://www.uu.se/en/linnaeus-garden/our-plants-and-attractions/our-plants/linnaeuss-floral-clock
26 I am grateful to Jesper Kårehed, Curator of the Linnaeus Garden at Uppsala University, for confirming this.
27 For Linnaeus's own intentions for the floral clock, and the first suggestion that it could be grown in a flower bed, see Karin Martinsson, *Linnés Blomsterur*, Prisma, Stockholm, 2003, pp. 18 and 35.
28 Ibid.
29 In 1777 the writer Hannah More (1745–1833) raised a satirical eyebrow at the fashion of the London ladies she had met at a party. 'I protest I hardly do them justice,' she wrote in a letter, 'when I pronounce that they had amongst them, on their heads, an acre and a half of shrubbery, besides slopes, grass plats, tulip beds, clumps of peonies, kitchen gardens and greenhouses.' (quoted in Charlotte M. Yonge, *Hannah More*, W. H. Allen & Co., London, 1888, pp. 26–7).
30 *Loves* is the second part of Erasmus Darwin's *The Botanic Garden*, and was published ahead of the first part.
31 Tristanne Connolly, 'Introduction', in *Erasmus Darwin's The Loves of the Plants*, Connolly, Elizabeth Bernath and Alana

Rigby (eds), *Romantic Circles Editions*, 2025, https://romantic-circles.org/editions/loves-plants

32. For a fascinating discussion of Linnaeus's sexual system and its subversive undercurrents, see Staffan Müller-Wille, 'Linnaeus and the love lives of plants', in Nick Hopwood, Rebecca Flemming and Lauren Kassell, *Reproduction: Antiquity to the Present Day*, Cambridge University Press, 2018, pp. 305–18.

33. In 1808 Rev. Samuel Goodenough wrote to a correspondent at the Linnean Society that '[a] literal translation of the first principles of Linnaean botany is enough to shock female modesty'. In William Withering's guide to native British plants using Linnaean classification, he explained that, '[f]rom an apprehension that Botany in an English dress would become a favourite amusement with the Ladies … it was thought proper to drop the sexual distinctions' (*A Botanical Arrangement of All the Vegetables Naturally Growing in Great Britain*, Vol. 1, T. Cadel and P. Elmsley/G. Robinson, London, 1776, p. v).

34. See Müller-Wille, 'Linnaeus', p. 309, and Connolly, 'Introduction'.

35. Erasmus Darwin, *The Loves of the Plants*, Canto I, lines 9–10.

36. Ibid., Canto IV, lines 15, 20, 28 and 29.

37. Quoted in Connolly, 'Introduction'.

38. Erasmus Darwin, *The Loves of the Plants*, Canto II, line 170. Also see note to line 165.

39. See Patricia Fara, *Sex, Science, & Serendipity*, Oxford University Press, 2012. Also Connolly, 'Introduction'.

40. Hemans' popular poem is quoted, for instance, in Anne Pratt, *Flowers and Their Associations*, Charles Knight and Co., London, 1840, p. 95; and Florence Caddy, *Through the Fields with Linnaeus: A Chapter in Swedish History*, Vol. 2, Longmans, London, 1887, p. 295.

41. These words by Tennyson were published in 1847. For me, they recall opening lines in Erasmus Darwin's *Loves*, such as, 'With secret sighs the Virgin Lily droops', and a line in another notable poem, Charlotte Smith's 'Horologe of the Fields' (1807), where Silene 'gives all her sweetness to the night'.

42. Caddy, *Linnaeus*, pp. 295–6.

43. See Kerner, *Natural History*, p. 215.

44 Another is that the plant's inner clock orchestrates the right time to pour effort into growing. Green plants store the energy they produce during the day and consume it to grow overnight.
45 See Antony N. Dodd et al., 'Plant circadian clocks increase photosynthesis, growth, survival, and competitive advantage', *Science*, Vol. 309, No. 5734, 2005, doi:10.1126/science.1115581
46 Jole Shackelford, *An Introduction to the History of Chronobiology*, Vol. 1, University of Pittsburgh Press, 2022, pp. 57 and 62.
47 Hagop S. Atamian et al., 'Circadian regulation of sunflower heliotropism, floral orientation, and pollinator visits', *Science*, Vol. 353, No. 6299, 2016, doi:10.1126/science.aaf9793
48 Jane J. Lee, 'How a Rooster Knows to Crow at Dawn', *National Geographic*, 19 March 2013, https://www.nationalgeographic.com/animals/article/130318-rooster-crow-circadian-clock-science
49 Danny Kessler, Celia Diezel and Ian T. Baldwin, 'Changing Pollinators as a Means of Escaping Herbivores', *Current Biology*, Vol. 10, No. 3, 2010, https://doi.org/10.1016/j.cub.2009.11.071
50 By which I mean, like the protagonist of Virginia Woolf's novel *Orlando: A Biography* (1928).
51 See '*Victoria cruziana*', Cambridge University Botanic Garden, https://www.botanic.cam.ac.uk/the-garden/plant-list/victoria-cruziana
52 See Kerner, *Natural History*, p. 215.
53 For influences on timing, see Wouter G. van Doorn and Chanattika Kamdee, 'Flower opening and closure: an update', *Journal of Experimental Botany*, Vol. 65, No. 20, 2014, p. 5750.
54 'Also,' Sylvain added, 'we grew the floral clock in a dry year, and the plants were quite stressed despite all our efforts. That might have been another reason why their timing shifted.'

As for variations in timing recorded by different studies of a particular species, Sylvain suspected that genetic diversity may play a part, in addition to environmental cues like day length and temperature.
55 For a compelling discussion of broader issues surrounding how we relate to more-than-human life, see Merlin Sheldrake, *Entangled Life: How Fungi Make Our Worlds, Change Our Minds, and Share Our Futures*, Vintage, 2021, pp. 45–6 and 73–4.

NOTES

3. WAULKING SONGS AND FURLONG-WAYS

1 This fictional opening scene and the following account of Hebridean labour songs draw deeply on the work of the Gaelic song specialist Margaret Fay Shaw, the folklorist John Lorne Campbell (her husband) and his collaborator Francis Collinson. This chapter is also greatly indebted to the Gaelic tradition-bearers Frances Dunlop and Fiona J. Mackenzie for their generous insights during our conversations in 2024 and 2025. Any errors are, of course, my own.

   The sources I have relied on most are:

   Margaret Shaw Campbell, 'Hunting folk songs in the Hebrides', *National Geographic*, February 1947, pp. 249–72.

   Margaret Fay Shaw, *Folksongs and Folklore of South Uist*, 3rd ed., Aberdeen University Press, 1986. (First published in 1955.)

   —, 'The folk songs of South Uist', a radio documentary for the BBC Third Programme, 1956, reproduced in the video *Folksongs of South Uist Broadcast – Margaret Fay Shaw's words read by Fiona J. Mackenzie*, 2015, https://www.youtube.com/watch?v=UQ7mMOyT8Gs

   Campbell and Collinson (eds), *Hebridean Folksongs I: A Collection of Waulking Songs Made by Donald MacCormick in Kilphedir in South Uist in the Year 1893*, Oxford at the Clarendon Press, 1969.

   —, *Hebridean Folksongs II: Waulking Songs from Barra, South Uist, Eriskay and Benbecula*, Oxford at the Clarendon Press, 1977.

   —, *Hebridean Folksongs III: Waulking Songs from Vatersay, Barra, South Uist, Eriskay and Benbecula*, Birlinn, Edinburgh, 2019. (First published in 1981.)

   Fiona J. Mackenzie, 'Bho mhoch gu dubh – from dawn to dusk', National Trust for Scotland, 9 July 2020, https://www.nts.org.uk/stories/bho-mhoch-gu-dubh-from-dawn-to-dusk

   Tobar an Dualchais/Kist o Riches, the online resource dedicated to audio recordings of Scotland's cultural heritage, tobarandualchais.co.uk

2 Quoted by Campbell and Collinson, *Folksongs III*, p. 6.
3 See Ted Gioia, *Work Songs*, Duke University Press, Durham and London, 2006.

4 See Ch. 2 (especially p. 27) in Marek Korczynski, Michael Pickering and Emma Robertson, *Rhythms of Labour: Music at Work in Britain*, Cambridge University Press, 2013. The authors argue that the 'work song literature' has been too narrowly focused on songs with functional purpose, to the exclusion of other cultures of singing at work in Britain (p. 33). Even the waulking songs and sea shanties – forms which seem 'to have a primarily functional value in the pacing, rhythm and coordination of labour' (p. 74) – 'had the dual dimension of stimulating and synchronising physical action, while also appealing to the emotions and imaginations of singers and listeners' (p. 82).

5 The lyrics here paraphrase part of the translation of the song, 'Gura Mise Tha Fo Éislein', given by Shaw in *Folksongs*, pp. 216–17.

6 Miss Goodrich Freer quoted in Campbell and Collinson, *Folksongs I*, p. 9.

7 Shaw, 'South Uist', BBC.

8 'Chunnaic mi an t-òg uasal' / 'I saw the well-born youth', Mary Morrison singing, in Campbell and Collinson, *Folksongs III*, pp. 118–19.

9 'Thug an latha gu dìle' / 'The day turned to downpour', Nan McKinnon singing, in Campbell and Collinson, *Folksongs III*, pp. 216–9 and 293–4. The story referred to is not known.

10 See Rona Wilkie, 'Performing Waulking Songs as an Emotional Practice in Gaelic Scotland', in Josephine Hoegaerts and Janice Schroeder (eds), *Ordinary Oralities: Everyday Voices in History*, De Gruyter Oldenbourg, 2023, p. 138.

11 For the tempi of waulking and clapping songs, see Campbell and Collinson, *Folksongs III*, pp. 314–16.

12 These words paraphrase a version of a song from Barra called 'Chan Eil Mi Gun Mhulad Orm', which Frances Dunlop translated for me from memory.

For the sexual politics expressed in the songs, see Wilkie, 'Waulking Songs', especially pp. 141–2.

13 Sgioba Luaidh Inbhirchluaidh (Inverclyde Waulking Group), waulk.org

14 The song is called, 'Chan Ann Leis An Fhuachd A Chailleadh An Duin' Agam'.
15 Goodrich Freer quoted in Campbell and Collinson, *Folksongs I*, p. 9.
16 Quoted in Campbell and Collinson, *Folksongs III*, pp. 6 and 7.
17 Shaw, *Folksongs*, p. 6.
18 Paul Nichols, 'Anatomy of a pop song', *M Magazine*, PRS for Music, 26 September 2017, https://www.prsformusic.com/m-magazine/how-to/anatomy-pop-song
19 Margaret Fay Shaw, *From the Alleghenies to the Hebrides: An Autobiography*, Birlinn, Edinburgh, 2008, p. 81.
20 Shaw, 'Hunting', p. 249.
21 Shaw, *Alleghenies*, p. 81.
22 Shaw, 'Hunting', p. 259. There was a widespread tradition of milkers blessing and encouraging the cow by singing her praises.
23 Shaw, *Folksongs*, p. 6.
24 Shaw, 'Folk songs' (BBC).
25 See Thomas Pennant quoted in Campbell and Collinson, *Folksongs I*, p. 4.
26 Annie Johnston, 'Brà brà bleith', Barra, 3 August 1950, track 25903, Tobar an Dualchais, https://www.tobarandualchais.co.uk/track/25903?l=en
27 See Korczynski, Pickering and Robertson, *Rhythms of Labour*.
28 My account of the knitters in this chapter leans heavily on Marie Hartley and Joan Ingilby, *The Old Hand-Knitters of the Dales*, The Dalesman, Clapham, via Lancaster, Yorkshire, 1969. (First published in 1951.)
29 Quoted in ibid., p. 47.
30 Ibid., p. 17.
31 As recorded and spelled in an account from Wensleydale (R.S.T., 'Knitting Song', *Notes and Queries*, Ser. 3, Vol. 4, London, July–September 1863, p. 205). Also see Hartley and Ingilby, *Hand-Knitters*, p. 16.
32 Hartley and Ingilby, *Hand-Knitters*, pp. 16 and 62.
33 This anonymous poem was 'said of old', according to Adam Sedgwick in *A Memorial to Cowgill Chapel* (1868, quoted in

Hartley and Ingilby, *Hand-Knitters*, p. 58). And it may not be a great exaggeration. See Walker's observations about a knitter called Slinger, quoted in *Hand-Knitters*, p. 47.

34 Hartley and Ingilby, *Hand-Knitters*, p. 16.
35 Quoted in ibid., p. 61.
36 Thomas Wright, *The Romance of the Lace Pillow*, H. H. Armstrong, Olney, Bucks, 1919, pp. 102 and 179.
37 Ibid., p. 106.
38 Ibid., pp. 179 and 193.
39 Ibid., pp. 180–1.
40 See Korczynski, Pickering and Robertson, *Rhythms*, p. 131.
41 Quoted in Hartley and Ingilby, *Hand-Knitters*, pp. 58–60. Also see Howitt's account, ibid., p. 63.
42 Hartley and Ingilby, *Hand-Knitters*, p. 16.
43 Recorded by Howitt, quoted in ibid., p. 62.
44 Ibid., pp. 65 and 67.
45 F. W. Moorman (ed.), *Yorkshire Dialect Poems (1673–1915) and Traditional Poems*, Sidgwick and Jackson, London, 1916, p. 53.
46 Hartley and Ingilby, *Hand-Knitters*, p. 28.
47 See the fifteenth-century recipe for azure paint, for example, in Mary Merrifield (ed.), *Original Treatises Dating from the XIIth to XVIIIth Centuries on the Arts of Painting*, Vol. 2, John Murray, London, 1849, p. 358.
48 See Gockerell, 'Telling Time', p. 142.
49 Quoted in Henry Notaker, *A History of Cookbooks: From Kitchen to Page Over Seven Centuries*, University of California Press, 2022, p. 121.
50 J. A. Vernon and T. E. McGill, 'Time estimations during sensory deprivation', *The Journal of General Psychology*, 69 (1), 1963 pp. 11–18, https://doi.org/10.1080/00221309.1963.9918425. In general, studies indicate that when external time cues are removed from participants in isolation experiments, that can significantly reduce their ability to orient themselves in time. For a discussion of the limited research in this area, see Virginie van Wassenhove, 'Temporal disorientations and distortions during isolation,' *Neuroscience & Biobehavioral Reviews*, Vol. 137, 2022, https://doi.org/10.1016/j.neubiorev.2022.104644.

51 Gockerell, 'Telling Time', p. 138.
52 'Pissing While, *N.*', *OED*, September 2024, https://doi.org/10.1093/OED/1088358304
53 For example, in *The Miller's Tale*.
54 'Fride Creme of Almaundys' in *Two Fifteenth-Century Cookery-Books: Harleian MS. 279 (ab 1430), & Harl. MS. 4016 (ab. 1450)*, Thomas Austin (ed.), Oxford University Press, London, 1964, p. 7, in Corpus of Middle English Prose and Verse, University of Michigan Library Digital Collections, https://name.umdl.umich.edu/CookBk
55 *The Treasure of Pore Men*, London, 1556, fol. xxx.
56 Thomas Glick, Steven J. Livesy and Faith Wallis (eds), *Medieval Science, Technology, and Medicine: An Encyclopedia*, Routledge, New York and London, 2005, p. 128.
57 *A Treatise on the Astrolabe*, 1.16.
58 For the probable distance measured by a mile for Chaucer, see Gerhard Dohrn-van Rossum, *History of the Hour: Clocks and Modern Temporal Orders*, Thomas Dunlap (trans.), University of Chicago Press, Chicago and London, 1996, p. 300.
59 'York Minster Masons' Ordinances, 1370' in Douglas Knoop and G. P. Jones, *The Mediæval Mason: An Economic History of English Stone Building in the Later Middle Ages and Early Modern Times*, Manchester University Press, 1933, pp. 248–9.
60 Other measures, such as 'half a mileway', are used in the ordinance too.
61 See, for example, 'The manner of the safe-keeping of the City, in the King's behalf', 1321, in Henry Thomas Riley (ed.), *Memorials of London and London life, in the XIIIth, XIVth, and XVth centuries*, Longmans, Green, and Co., London, 1868, pp. 142–4.
62 For an important analysis focused on the city of York in this period, see Chris Humphrey, 'Time and Urban Culture in Late Medieval England', in Humphrey and W. Mark Ormrod (eds), *Time in the Medieval World*, York Medieval Press, 2001.
63 Barbara Tuchman quoted in Ken Mondschein and Denis Casey's helpful overview of medieval philosophies of time and systems of measurement, 'Time and Timekeeping', in Albrecht Classen (ed.), *Handbook of Medieval Culture: Fundamental*

Aspects and Conditions of the European Middle Ages, Vol. 3, Walter de Gruyter GmbH, Berlin, 2015, p. 1678.

64 Geoffrey Chaucer, *Boece*, Book 1, Metre 3, line 6.
65 See Rossum, *The Hour*, pp. 300–1 and 304.
66 Quoted in Gioia, *Work Songs*, p. 114.
67 Ibid., p. 113.
68 This was an intensification of existing practices (see Rossum, *The Hour*, pp. 318–21). By the late eighteenth century in industrial settings, Gioia writes, '[t]he fundamental tempo of work, once demarcated and paced by song, would now be established by the much different sound of machines in motion' (*Work Songs*, p. 103).
69 Campbell and Collinson, *Folksongs I*, p. 16. His fuller argument may sound both idealising and overly deterministic and dismissive. Yet there is an undeniable appeal, I think, to the creative communal life he describes.
70 Shaw remarked in her radio script that South Uist was 'one of the few places left where genuine waulking' was still performed. Also see Campbell and Collinson, *Hebridean Folksongs II*, p. 2, and *III*, p. 5.
71 In general, in the factories, singing together while working had been silenced by order or by mechanical noise. However, in the 1940s in Britain, efforts to offset the dehumanising effects of Taylorist production while boosting efficiency included piping rhythmic music to the factory floor. See chapters 7–9 in Korczynski, Pickering and Robertson, *Rhythms of Labour*.
72 Shaw, 'South Uist', BBC. My thanks to Fiona for sharing this quotation from Shaw's original script, which is held at the Canna House Archive, National Trust for Scotland.
73 See Fiona J. Mackenzie, *The Cadence of a Song: The Life of Margaret Fay Shaw*, Birlinn, Edinburgh, 2025.
74 See Wilkie, 'Waulking Songs'.
75 See Ian Sample, '"A mega-mechanism for bonding": why singing together does us good', *Guardian*, 15 December 2023, https://www.theguardian.com/science/2023/dec/15/a-mega-mechanism-for-bonding-why-singing-together-does-us-good

NOTES

76 Quoted in Andrea, 'Waulking the Tweed', Skye Weavers, 24 May 2024, https://www.skyeweavers.co.uk/blog/waulking-the-tweed

77 During the Clearances in the Highlands and western islands of Scotland in the eighteenth and nineteenth centuries, small farming communities were evicted and displaced by their landlords, often with devastating consequences.

78 Auchindrain is now a museum.

4. SCRATCH DIALS AND STICK DIALS

1 A fragment of what is believed to be a Roman sundial, dating from between the second and fourth centuries CE, was found at Housesteads Roman Fort on Hadrian's Wall and is now displayed at Chesters Roman Fort.

2 David Scott and Mike Cowham, *Time Reckoning in the Medieval World: A Study of Anglo-Saxon and Early Norman Sundials*, BSS Monograph No. 8, British Sundial Society, 2010, p.1 – citing Mario Arnaldi, *The Ancient Sundials of Ireland*, British Sundial Society, 2000 – and pp. 8–9.

3 The 'equal' or 'equinoctial' hour had been used for millennia in specialist settings. The spread of mechanical clock bells (and hourglasses) in late-medieval Europe brought the equal hour into everyday life. See Bonnie Blackburn and Leofranc Holford-Strevens, *The Oxford Companion to the Year*, Oxford University Press, 1999, p. 662.

4 The canonical hours are the daily periods of Christian prayer and reflection. In the early years, Prime was celebrated at sunrise. The next three canonical hours – Terce (around mid-morning), Sext (near midday) and None (around mid-afternoon) – were tied to the third, sixth and ninth Roman daylight hours. Vespers was sung around sunset, Compline at the descent into darkness, Nocturne around midnight and, as the night drew to a close, Matins began the new liturgical day. See Mario Arnaldi, 'Time Reckoning in the Medieval World', in Anthony Turner, James Nye and Jonathan Betts (eds), *A General History of Horology*, Oxford University Press, 2022, p. 102.

5 For the flexibility of times of worship, and medieval time structures more generally, see Mario Arnaldi, 'Intuitive geometries of time perception, from Antiquity to the Middle Ages', in Laura Farroni, Manuela Incerti and Alessandra Pagliano (eds), *Rappresentare il tempo: Architettura geometria e astronomia*, Libreriauniversitaria.it Edizioni, Padua, 2022, p. 197; W. Rothwell, 'The Hours of Day in Medieval French', *French Studies*, Vol. 13, No. 3, July 1959, p. 241; and Thomas Glick, Steven J. Livesy and Faith Wallis (eds), *Medieval Science, Technology, and Medicine: An Encyclopedia*, Routledge, New York and London, 2005, pp. 127–30.

6 See Gillian Adler and Paul Strohm, *Alle Thyng Hath Tyme: Time and Medieval Life*, Reaktion Books, London, 2023. Also Rothwell, 'Hours', especially pp. 247–9.

7 See Jérôme Bonnin, 'Time Measurement in Antiquity' in Turner, Nye and Betts, *Horology*, especially p. 17, and Arnaldi, 'Intuitive Geometries', p. 188.

8 The Mass Dial Register.

9 In 1996 the British Sundial Society Mass Dial Group made a tour of scratch dials in this area, focusing on All Saints church at Woolstone, led by Frank Poller. Margaret Stanier's account of the trip ('Mass Dials – in the Vale of the White Horse', *The British Sundial Society Bulletin (BSSB)*, No. 97.2, April 1997, p. 10) caught my eye and drew me to Woolstone. I've also consulted Poller's survey, *Medieval scratch dials on Vale of White Horse churches*, 3rd issue, the author, Oxon., 1996.

10 I had been hoping to find this dial, having read about it in Stanier's report ('Mass Dials').

11 Even under clear skies, the shadow of the gnomon wouldn't show up to mark time on the dial all through the day, every day. In the northern hemisphere, in the summer half of the year between the spring and autumn equinoxes, the Sun rises north of east and sets north of west. That means it will be later in the morning by the time sunbeams hit the dial on a south-facing wall. Likewise, as the Sun comes closer to setting, its rays will no longer illuminate the wall.

12 David Scott, 'Sundials in Anglo-Saxon England, Part 4, The Late Period – Aldbrough and Orpington', *BSSB*, Vol. 12 (i),

February 2000, p. 38, and 'The Perception of Time in Anglo-Saxon England', *BSSB*, Vol. 22 (iv), December 2010, p. 37.
13 Mario's 'Moon Man' (2013) was inspired by Cyrano de Bergerac's seventeenth-century novel, *The Voyage to the Moon*, where lunarians measure time by the shadow of their nose falling on their teeth. Just so, the shadow of the Moon Man's beaky nose moving over the numbered teeth in his great grin tells the clock hour precisely.
14 Arnaldi, 'Time Reckoning', p. 110.
15 See Arnaldi, 'Time Reckoning', pp. 102–12.
16 See Ben Jones, 'Some Scratch Dials and Other Graffiti', *BSSB* 35 (iv) December 2023, p. 32.
17 Tony Wood, 'Scratch Dials in Gloucestershire', *Gloucestershire History*, No. 21, 2007, p. 16.
18 Stanier, 'Mass Dials', p. 10.
19 Poller, *Scratch dials*, p. 15.
20 For the kind of disputes about timing that might arise between medieval laity and priests, see Nicholas Orme, *Going to Church in Medieval England*, Yale University Press, New Haven and London, 2021, pp. 228–9.
21 Arnaldi, 'Time Reckoning', p. 102.
22 The shift does not seem to have happened everywhere all at once. For historical examples, see 'Noon, *N.*, Sense 1. a–b; 2.a', *OED*, June 2025, https://doi.org/10.1093/OED/2667346547
23 See 'Noon, *N.*, Etymology', ibid., https://doi.org/10.1093/OED/1040985504. Also Rothwell, 'Hours of the Day', pp. 244–5. For a more sceptical discussion, see David S. Landes, *Revolutions in Time: Clocks and the Making of the Modern World*, The Belknap Press of Harvard University Press, Cambridge, Mass., Rev. Ed., 2000, pp. 439–40, note 25.
24 See Poller, *Scratch dials*, pp. 15–6. Etched into the stone surround of a north window at the eastern end of the church is a large neat circle divided by lines that Poller speculated had been a scratch dial. Future analysis may make its purpose and history clearer too.
25 After collecting more than 3,500 examples of scratch dials, the British Sundial Society confirms the pattern and placement of many of them leaves little doubt they were rudimentary

devices for marking events by shadows. Some dial-like forms, however, may be symbolic motifs or other forms of graffiti or device. Others may be grooves and holes made incidentally. See Karlheinz Schaldach, 'Circular Objects', *BSSB*, Vol. 36 (ii), June 2024. Also Matthew Champion, *Medieval Graffiti: The Lost Voices of England's Churches*, Ebury Press, London, 2015, pp. 158–9.

26 See Champion, *Graffiti*, especially pp. 5, 25–8, and 75–8. More properly, they are known as apotropaic or ritual protection marks. Some of the marks were made for other purposes, for example, as memorials.

27 For the possible symbolism of circular marks, see Schaldach, 'Circular Objects', p. 14.; Champion, *Graffiti*, pp. 75–8; and Jones, 'Scratch Dials'.

28 This gulf is powerfully expressed in Champion, *Medieval Graffiti*.

29 When I shared the photograph with Mario, he suspected the orientation of the wall might help explain the placement of the lines. But in this case, he concluded after analysing the photo of the dial mathematically, that doesn't seem to be the solution to the puzzle of its pattern. Further on-site investigation is needed.

For a discussion of attempts to improve the accuracy of medieval dials, see Mario Arnaldi, 'The medieval Rule of Erfurt written in a codex that belonged to Fra Giocondo of Verona', *BSSB*, Vol. 35 (iii), September 2023, pp. 2–9.

30 Wood, 'Gloucestershire', p. 16.

31 Hilaire Belloc, 'On a Sundial', 1938.

32 For the social and cultural impact of the nationalisation then globalisation of Western time standards, especially Greenwich Mean Time, see Vanessa Ogle, *The Global Transformation of Time, 1870–1950*, Harvard University Press, 2015, and David Rooney, *About Time: A History of Civilization in Twelve Clocks*, Viking, 2021.

33 By the early 1700s, the Sun's role as the source of 'true' time had been challenged by new developments in scientific time measurement, which laid bare how the Sun appears to take a little longer or shorter than twenty-four hours to return to

its midday position. However, the Sun continued to be the reference for regulating clocks. See Anthony Turner, 'The Eclipse of the Sun: Sun-dials, Clocks and Natural Time in the Late Seventeenth Century', *Early Science and Medicine*, Vol. 20, No. 2, 2015, pp. 177 and 182–3.

34 Arnaldi, 'Time Reckoning', p. 114.
35 See Mario Arnaldi and Karlheinz Schaldach, 'A Roman Cylinder Dial: Witness to a Forgotten Tradition', *Journal for the History of Astronomy*, xxviii, 1997, pp. 107–17.
36 'Evening Quatrains'.
37 See Arnaldi, 'Time Reckoning', p. 113.
38 He guesses the time by a combination of signs, including the length of his shadow. See Adler and Strohm, *Alle Thyng*, pp. 11–12.
39 Jonas Vaiškūnas, 'The Observation of Celestial Bodies and Time Counting in the Lithuanian Folk Culture', *Time and Astronomy in Past Cultures*, Arkadiusz Sołtysiak (ed.), Gorgias Press, USA, 2009, pp. 169–70 and 188.
40 Memoir 1857/1970–1, Ethnological Collection of the National Museum of Iceland, sarpur.is
41 See Mario Arnaldi, *Orologi solari a Taggia – antiche conoscenze del tempo tra scienza e costume*, Comune di Taggia, 1996, pp. 66–8.
42 See the photograph by Marco Rech in Gabriele Vanin, *Le meridiane bellunesi, quaderno n.9, Comunità Montana Feltrina – Centro per la documentazione della cultura popolare*, Libreria Pilotto Editrice, Feltre, 1991, p. 23.
43 Geoffrey Grigson, *The Shell Country Alphabet: The Classic Guide to the English Countryside*, Penguin, 2010, p. 380.
44 Paul Cavill, *A New Dictionary of English Field-Names*, English Place-Name Society, Nottingham, 2018, pp. 110–11.
45 *Henry VI*, Part 3, Act 2, Scene 5.
46 Edward Lovett, 'The Simple Sundial of the South-Down Shepherds', 1909, in Adelaide A. J. Gosset (ed.), *Shepherds of Britain: Scenes from Shepherd Life Past and Present from The Best Authorities*, Constable and Co., London, 1911, pp. 272–4.
47 Lovett reports that one of his interviewees, William Aylward, 'invented a turf dial, which for its simplicity beats all others …

He thus describes this method on the Downs: "On a sunny day with the south wind we could hear the cathedral clock strike, and if the sun was shining we used to fix a short stick upright in the ground and cut a ridge in the turf where the shadow fell, and so on at each hour; and on other days, when the striking of the clock could not be heard, but the sun shining, the dial was ready for use."' ('Simple Sundial', p. 274, note 1). Aylward may have discovered this technique independently. However, a similar method was used in Lithuania – see Vaiškūnas, 'Celestial Bodies', p. 170.

48 Take, for example, the splendid late-medieval astronomical clock at Wells Cathedral. The dial is divided into twenty-four hours (in two sets of twelve hours), with midday at the top and midnight below. The hour hand ferries a miniature golden sun around the dial.

49 See Turner, 'Eclipse', p. 169.

50 Lovett, 'Simple Sundial', p. 273.

51 See Joshua Pollard, 'The Uffington White Horse geoglyph as sun-horse', *Antiquity* 91, 2017, p. 407, https://doi.org/10.15184/AQY.2016.269

52 Lovett, 'Simple Sundial', p. 274, note 1.

5. DAYMARKS

1 I have translated these place names with the timing I think we would give them in English today. See page p. 120.

2 To search for daymark names, see the National Science Institute of Iceland map viewer, Kortasja.lmi.is. For further resources, see the place name collection at the Árni Magnússon Institute for Icelandic studies, Nafniðis (nafnid.is).

3 As astronomers warn, never look directly at or near the Sun with the naked eye or through sunglasses, binoculars or any kind of optical instrument that has not been fitted with the right filters. Serious and permanent eye damage can occur in an instant.

4 A small population of Christian Gaels are believed to have made some kind of life in Iceland around 800 CE. However, the founding Norse-led population of Iceland are thought to have come from Scandinavia, Ireland and Scotland several decades later.

NOTES

5   Andrew Dennis, Peter Foote and Richard Perkins (trans.), *Laws of Early Iceland: Grágás I*, University of Manitoba Press, Winnipeg, 2006, p. 57.
6   Ibid., p. 59.
7   Kristian Kålund, *Bidrag til en historisk-topografisk Beskrivelse af Island*, Vol. 1, Gyldendalske Boghandel, Copenhagen, 1877, pp. 111–13.
8   People would also have told time by shadows.
9   The trio is based on actual hills on the map named Heimasti-Nónhóll, Mið-Nónhóll and Syðsti-Nónhóll.
10 Thorsteinn Vilhjálmsson, 'The Subarctic as a Sundial', in C. Esteban and J. A. Belmonte (eds), *Oxford IV and SEAC 99: Astronomy and Cultural Diversity*, OACIMC, Tenerife, 2000, p. 213.
11 See Gísli Sigurðsson, 'Foreword', in Pétur Halldórsson, *The Measure of the Cosmos: Deciphering the Imagery of Icelandic Myth*, Iconea Publications, London, 2022, pp. 10–11.
12 Finn Magnusen, 'On the ancient Scandinavians' division of the times of the day', *Mémoires de la Société Royale des Antiquaires du Nord*, Vol. 1, 1839, pp. 165–92.
13 See Sigrun Kristjansdóttir, 'Afmörkun timans', in Árni Bjornsson et al. (eds), *Hlutavelta tímans: menningararfur á Þjóðminjasafni*, National Museum of Iceland, 2004, pp. 324–31.
14 Birna Lárusdóttir, 'Jöfnubáðu-örnefni og vangaveltur um eyktamörk', Árnastofnun, 1 January 2019, https://www.arnastofnun.is/is/utgafa-og-gagnasofn/pistlar/jofnubadu-ornefni-og-vangaveltur-um-eyktamork
15 'Methúsalem Methúsalemsson (1959)', Bustarfell, nafnið.is, https://nafnid.is/baer/936
16 My thanks to Johan Anton Wikander for telling me about these location names.
    'Undorn' is an old word for a time of day that varied between regions and eras. See the entries for both Undersåker and Undersvik in the Swedish Place Name Dictionary at isof.se, the website of The Institute for Language and Folklore, Sweden, https://www.isof.se/namn/ortnamn/sol/ortnamnslexikon/ucklum-uvered

17 David Raven, 'Noon Stone', Yorkshire (West), *The Megalithic Portal*, 26 November 2004, https://www.megalithic.co.uk/article.php?sid=7398
18 Gockerell, 'Telling Time', p. 135.
19 As Thorsteinn Vilhjálmsson suggested to me in 2024, when clocks were introduced, possibly they were used to determine the *eyktamörk* with more precision.
20 Memoir 1857/1970–1, Ethnological Collection of the National Museum of Iceland, sarpur.is
21 See Þorsteinn Sæmundsson, 'Um tímareikning á Íslandi', Almanak Háskóla Íslands, Institute of Natural Sciences, University of Iceland, 1988, http://almanak.hi.is/klukkan.html
22 The Icelandic words *dagsmark* and *eyktamark* are used interchangeably for a physical object in the landscape that marks time by the Sun. That seems to be the limit of the meaning of *eyktamark*. Whereas *dagsmark* appears to have a broader range of connotations. On the one hand, *dagsmark* could mean a certain time of day or night in itself without reference to the horizon. On the other, it seems likely that anything that helped you tell time – a sundial, a certain star culminating at midnight, an hourglass – may have been considered a *dagsmark*. See the entries for *dagsmark* and *eyktamark* at the Árni Magnússon Institute's language database, málið.is (malid.is).
23 Albert E. Waugh, *Sundials: Their Theory and Construction*, Dover Publications, New York, 1973, p. 18.
24 P. Gemsege, *The Gentleman's Magazine*, Vol. 29, E. Cave, London, 1759, p. 16.
25 Waugh, *Sundials*, p. 20.
26 The map shows the Stonehenge monument as it was in around 2500 BCE. It is reproduced in the likeness of fig. 2.7 in Duncan Garrow and Neil Wilkin, *The World of Stonehenge*, British Museum, London, 2022, p. 81. Also see pp. 77–85.
27 See Joshua Pollard, 'The Uffington White Horse geoglyph as sun-horse', Antiquity 91, 2017.
28 See Garrow and Wilkin, *Stonehenge*, pp. 133–44.
29 Liliane Lijn, 'Light Pyramid', 2012.
30 In his essay, 'The Subarctic as a Sundial'.

NOTES

31 Another of Stuart Mosscrop's major contributions to CMK was to co-design the original shopping centre, which opened beside Midsummer Boulevard in 1979. In recent years, this highly regarded (and Grade II-listed) building has undergone a major extension that cuts right across the boulevard – to Stuart's understandable dismay. You have to pass through the shopping centre now if you want to stay on the route. Moreover, it means sunbeams no longer flow uninhibited along this part of Midsummer Boulevard. The light is interrupted. And with the light, the time.

   An open grid allows for flow and adaptation; it is a means, Stuart explained, of making provision for unknown needs or uses beyond the present. When I asked what is lost by blocking the main artery of CMK's grid, his answer was emphatic: 'A chunk of the future!'

32 He explains, 'according to right reckoning the sun will pass through each division in three hours of the day' (*The King's Mirror (Speculum Regale – Konungs skuggsjá)*, Laurence Marcellus Larson (trans.), American-Scandinavian Foundation, New York; Oxford University Press, London, 1917, p. 93).

33 Ibid., p. 83.

34 Lárusdóttir, 'Jöfnubáðu-örnefni'.

6. THE GLOAMING AND THE DIMPSE

1 See 'Gloaming, *N.*, Etymology', *OED*, March 2025, https://doi.org/10.1093/OED/6685724759

2 My influences here include Steven Connor's 'A Certain Slant of Light', the script for a sound-essay on *Nightwaves*, BBC Radio 3, 31 October 2003, https://www.stevenconnor.com/twilight/; and James Harbeck's blog, 'Gloaming', *Sesquiotica*, 1 September 2019, https://sesquiotic.com/2019/09/01/gloaming/#comments

3 There are other factors at play in how we see colour. In this case, the daytime sky appears blue partly because our eyes are more sensitive to blue light. As experts have repeated to me, colour perception involves a very complex combination of biology, psychology and physics.

4 For an accessible explanation that clarifies common misunderstandings, see Stephen F. Corfidi, 'The Colors of

Sunset and Twilight', NOAA/NWS Storm Prediction Center, Norman, OK, 2014, https://www.spc.noaa.gov/publications/corfidi/sunset/

5 See Marcel Minnaert, *Light and Color in the Outdoors*, Len Seymour (trans.), Springer-Verlag, New York, 1993, pp. 292 and 306.

6 *The Nautical Almanac and Astronomical Ephemeris for the Year 1937 for the Meridian of the Royal Observatory at Greenwich*, His Majesty's Stationery Office, London, 1936, especially pp. iii and 702.

7 See Donald H. Sadler, 'A Personal History of H. M. Nautical Almanac Office, 30 October 1930–18 February 1972', George A. Wilkins (ed.), United Kingdom Hydrographic Office, 2008, p. 28.

8 In the age of GPS (GNSS), astro-navigation is still required and performed at sea as a backup in case of system failure, jamming or spoofing, or even disruption from space weather.

9 The first line of Chaucer's *A Treatise on the Astrolabe* is addressed to '[l]yte Lowys my sone'. He refers to 18° in Part 2, Canon 6. For his sources, see J. D. North, *Chaucer's Universe*, Clarendon Press, Oxford, 1988, p. 60. They are thought to include a compilation of treatises, mostly from the Islamic world, that were formerly attributed to the Persian Jewish scholar Masha'allah (Māshā'allāh ibn Atharī, c. 740–815 CE). My thanks to the historian of science Taha Yasin Arslan for his insights on this.

10 For an insider's account of how the *Nautical Almanac*'s subdivisions of twilight were defined and named, see Sadler, 'A Personal History', pp. 27–8. The project was led by Superintendent L. J. Comrie, who 'introduced two innovations: the concepts of civil and nautical twilight', according to Sadler (ibid.). 'If I remember correctly', he added, Comrie sought widely for ideas for their names (ibid.). Presumably Sadler was unaware that 'civil twilight' was coined long before the 1930s. (See 'Civil Twilight & Calculating the Day', *Wordplay: Science and Nature*, Merriam-Webster.com, undated, https://www.merriam-webster.com/wordplay/civil-twilight-word-history.) However, according to the *Oxford English Dictionary*,

the earliest evidence of 'nautical twilight' is indeed in *Nautical Almanac 1937* ('Nautical Twilight, N.', Factsheet, *OED*, September 2025, https://doi.org/10.1093/OED/4183352389). Future research may make the picture clearer. For the wider context, see Bernard D. Yallop and Catherine Y. Hohenkerk, 'The Almanacs in the 20th Century: Computers and Applications', in P. Kenneth Seidelmann and Catherine Y. Hohenkerk (ed.s), *The History of Celestial Navigation: The Rise of the Royal Observatory and Nautical Almanacs*, Springer Nature Switzerland AG, 2020, especially pp. 215–6.

11 *Nautical Almanac 1937*, p. 702.
12 This is from lines 45–6 in 'The Second Nun's Prologue', *Canterbury Tales*, following the translation given at Harvard's Geoffrey Chaucer Website, Canterbury Tales, Text and Translations, 8.1 The Second Nun's Prologue and Tale, https://chaucer.fas.harvard.edu/pages/second-nuns-prologue-and-tale
13 See, for instance, 'The manner of the safe-keeping of the City', 1321, Henry Thomas Riley (ed.), *Memorials of London and London life, in the XIIIth, XIVth, and XVth centuries*, Longmans, Green, and Co., London, 1868, p. 143.
14 Marcel Minnaert, *De natuurkunde van't vrije veld. Deel I: Licht en kleur in het landschap*, W. J. Thieme, Zutphen, 1937. As I write, I have two English-language editions to hand: *Light & Colour in the open air*, H. M. Kremer-Priest (trans.); K. E. Brian Kay (Rev.), G. Bell and Sons Ltd, London, 1959, and *Light and Color in the Outdoors*, 1993. On one occasion, noted below, I quote from the former translation when the latter tones down the poetry.
15 Minnaert, *Color* (1993), p. 294 and fig. 169.
16 This observation echoes a conversation in Hardy's earlier novel, *A Pair of Blue Eyes*. See page 7 in this book.
17 Fig. 170, 'Concise table showing the development of the different twilight phenomena', Minnaert, *Color* (1993), p. 299.
18 Ibid., p. 296.
19 Ibid., p. 292.
20 Minnaert, *Colour* (1959), p. v.
21 Minnaert, *Color* (1993), p. x.
22 Ibid., p. 297.

23 Kevin Lynch, *What Time Is This Place?*, Massachusetts Institute of Technology, 1972, p. 69.
24 Ibid., p. 148.
25 Ibid., p. 70.
26 The artwork was commissioned by Art on the Underground for Transport for London. (Years earlier I was curator for Art on the Underground but was not involved with this project.)
27 For a very old reference to forecasting by red skies, see Gospel of Matthew 16:2.
28 'Red sky at night and other weather lore', Met Office, https://weather.metoffice.gov.uk/learn-about/weather /how-weather-works/red-sky-at-night
29 Daniel Milligan, 'Ten words and phrases that prove you're Somerset born and bred', *Somerset County Gazette*, 17 February 2024, https://www.somersetcountygazette.co.uk/news/11011425 .ten-words-and-phrases-that-prove-youre-somerset-born -and-bred
30 Minnaert, *Color* (1993), p. 297.
31 Ibid.
32 These words are from *The Nautical Almanac and Astronomical Ephemeris for the Year 1942*, His Majesty's Stationery Office, London, 1941, pp. 508–9.
33 Jérôme Lalande, *Voyage d'un François en Italie*, Vol. 1, Paris, 1769, pp. xxxij–xxxiij.
34 In the fourteenth century, the first public clocks installed in Italian cities started the count of twenty-four equal hours at *sunset*. From the sixteenth century, that timing moved half an hour later to the bell for the Angelus prayer – that is, as Lalande reported to his French readers, when it's starting to get dark. See Mario Arnaldi, 'Italian Hours: Origin and Decline of One of the Most Important Time-Systems of the Past, Part 1', *BSSB*, Vol. 35 (iv), December 2023, and 'Part 2', *BSSB*, Vol. 36 (i), March 2024.
35 Robert Forby, *The Vocabulary of East Anglia*, Vol. 1, J. B. Nichols and Son, London, 1830, p. 89.
36 Thomas Miller, *Rural Sketches*, John Van Voorst, London, 1839, p. 118.
37 See Brigitte Steger, 'Landschaften der Zeit: Tages- und Nachtstunden im vormodernen Japan', Österreichische

NOTES

Forschungsgemeinschaft (eds), *Zeit in den Wissenschaften*, Böhlau, Vienna, 2016, pp. 59–60. The method marks when monks would have rung the bell for the sixth hour of morning. Strictly speaking, this was the start of the daylight but not the calendar day, Brigitte Steger explained to me. 'In most Buddhist traditions,' she said, 'the start of the day was in the early evening.'

38 See W. Rothwell, 'The Hours of Day in Medieval French', *French Studies*, Vol. 13, No. 3, July 1959, p. 243.

39 The day begins in the evening in several cultures, including Jewish and ancient Greek traditions. In Christian traditions, certain feasts and vigils begin on the evening before the significant day, for example, on Christmas Eve.

40 Isha follows Maghrib when the last of the light has faded. Fajr begins at first light. Zuhr when the Sun passes its highest point above you in the middle of the day (noon). Religious authorities vary in their interpretation of timings, however. For some, Asr begins when the shadow of an object has grown after noon by a span equal to its length. For others, Asr starts when the shadow after noon has grown by *twice* the length of the object. There are different opinions, too, about when Isha begins and ends.

41 The dates when the planet Venus appears in our sky change from one year to the next. Venus shows up as the evening star for a few months, disappears for a stretch, then returns as the morning star for another few months, before disappearing again and repeating the cycle.

42 Bede, *On the Nature of Things and On Times*, Calvin B. Kendall and Faith Wallis (trans.), Liverpool University Press, 2010, p. 108.

43 See 'cwyld-seten' in Joseph Bosworth, *An Anglo-Saxon Dictionary Online*, Thomas Northcote Toller, Christ Sean, and Ondřej Tichy (eds), Charles University, Prague, 2014, https://bosworthtoller.com/7029; also Frederick Tupper, Jr., 'Anglo-Saxon Dæg-Mæl', *Publications of the Modern Language Association*, Vol. 10, No. 2, 1895, p. 186.

44 Mishnah Berakhot 1:2.

45 Michal Vik and Renzo Shamey, 'Purkyně, Jan Evangelista', *Encyclopedia of Color Science and Technology*, Springer, New York, 2015, doi 10.1007/978-3-642-27851-8_342-1

46 Nicholas J. Wade, Josef Brožek and Jiří Hoskovec, *Purkinje's Vision: The Dawning of Neuroscience*, Psychology Press, New York, 2001, pp. 12–3.
47 C. T. Onions (ed.), *The Oxford Dictionary of English Etymology*, Oxford Univerity Press, 1994, p. 246.
48 Nicholas Breton, *Fantasticks: A perpetual Prognostication*, Francis Williams, London, 1626, p. 13.
49 My mind is on the cold North: see 'Kveld-riða' in the online Cleasby & Vigfusson – Old Norse Dictionary, https://cleasby-vigfusson-dictionary.vercel.app/word/kveld-rida
50 *Egeria: Diary of a Pilgrimage*, George E. Gingras (trans.), Newman Press, United Kingdom, 1970, p. 109.
51 Gerhard Dohrn-van Rossum, *History of the Hour: Clocks and Modern Temporal Orders*, Thomas Dunlap (trans.), University of Chicago Press, Chicago and London, 1996, p. 293.
52 William Ellis, *Polynesian Researches*, Vol. 1, Fisher, Son, & Jackson, London, 1831, p. 89.
53 *Cinnlae Amhlaoibh Uí Shúileabháin: The Diary of Humphrey O'Sullivan*, Part 2, Michael McGrath (ed.), Irish Texts Society, London, 1936, pp. 354–7. My gratitude to Kevin Murray and Geoffrey Keating for their expert help with this.
54 See Minnaert, *Color* (1993), p. 297.
55 On this cloudy evening, Aldebaran and Betelgeuse are the only visible objects bright enough to trigger our cone cells and activate colour vision.

7. STAR CLOCKS

1 My thanks to Farley Chase for sharing the story that inspired this reflection.
2 '"The Bear"' (without a modifier) refers to Ursa Major' ('Bear, *N.* (1), Sense I.4', *OED*, June 2025, https://doi.org/10.1093/OED/2895081447). 'Charles's Wain' is the figure formed by the seven brightest stars in Ursa Major ('Charles's Wain, *N.*', *OED*, June 2024, https://doi.org/10.1093/OED/8370300252).
3 The similarity of Hardy's descriptions at midwinter and in February implies he is referring to the same group of stars. Yet

the February phrase – 'at a right angle with the Pole Star' – is ambiguous.

I'm strongly inclined to believe Hardy means to repeat the midwinter image where 'the Bear had swung round [the North Star] outwardly to the east, till he was now at a right angle with the meridian.' Crucially, Ursa Major is indeed to the east of the North Star *both* at midnight on midwinter and at 9 p.m. (local apparent time) in February.

A more precise method would be to track time by the Little Bear (Ursa Minor), especially its two Guardian stars close to Polaris. Or by the two Pointer stars on the outer scoop of the Plough/Big Dipper which point to Polaris. But the huge body of the Plough is much more locatable to me on an average night in London, so my focus is on learning to tell approximate time by its rotation.

I'm grateful to Thomas Hockey and Fabio Silva for helping me think this through.

4 The editions I refer to in this chapter are *Half-Hours with the Stars*, Robert Hardwicke, London, 1869; and *A New Star Atlas ... In Twelve Circular Maps*, 2nd ed., Longmans, Green, and Co., London, 1872. These aren't the only books by Proctor that Hardy or his readers could have consulted to tell the time by the stars.
5 See Pamela Gossin, *Thomas Hardy's Novel Universe: Astronomy, Cosmology, and Gender in the Post-Darwinian World*, Routledge, 2007, especially chapters 4 and 5.
6 Richard Proctor, *Easy Star Lessons*, Chatto & Windus, London, 1881, p. 19.
7 *Half-Hours*, p. 9.
8 Ibid.
9 When Orion sets at this time of year, it is already broad daylight. That means we won't actually be able to see it disappear behind the western skyline.
10 See E. G. R. Taylor, *The Haven-Finding Art: A History of Navigation from Odysseus to Captain Cook*, Hollis & Carter, London, 1958, pp. 145–8.
11 A more complicated version of this map is given in fig. 7 in *A New Star Atlas*. Proctor continued to hone his explanations for the novice. After Hardy's novel was published, Proctor

published *Easy Star Lessons* (1881) with simpler charts, including a 'Star Clock' (see List of Illustrations and fig. 3). The two diagrams in this chapter showing how to tell time by the circumpolar stars are based on that Star Clock by Proctor, with the Guardians removed so that only the North Star and the Plough or Big Dipper remain. They were redrawn for me by the astronomer John A. Paice in 2025, who altered the position of the stars a little to make the diagrams accurate for our era.

12 'The shepherd's clock' ('horloge des bergiers'), *Le compot et calendrier des Bergiers*, 2nd ed., Guy Marchant, Paris, 1491, digitised at Les Bibliothèques Virtuelles Humanistes, University of Tours (https://www.bvh.univ-tours.fr/Consult/consult.asp?numfiche=905&numtable=B180336101%5FINC166&mode=1&ecran=0&index=102), f. h1 Ill. and f. h1v fig.

13 The first 'Shepherd's Calendar' was intended for a wealthy urban audience rather than the shepherds it appears to address, although versions of it were later peddled to rural readers. Various editions of the Shepherd's Calendar were produced in English and other languages. See Vincent Picard, 'Le Grand calendrier et compost des bergers', Dicopathe blog, 9 July 2022, https://www.dicopathe.com/le-grand-calendrier-et-compost-des-bergers/

14 For accounts of people in Lithuania in the nineteenth and twentieth centuries telling time by the stars, see Jonas Vaiškūnas, 'The Observation of Celestial Bodies and Time Counting in the Lithuanian Folk Culture', *Time and Astronomy in Past Cultures*, Arkadiusz Sołtysiak (ed.), Gorgias Press, USA, 2009, pp. 176–88. For accounts collected in Russia from the 1960s to the 1980s, see S. S. Petriashin, 'Rural diurnal timing practices based on stars observation', *Vestnik of Saint Petersburg University. History*, Vol. 62, No. 4, 2017, pp. 865–77. We will pick up on some of these practices later in the chapter.

15 Memoir 27/1976-3-27, Ethnological Collection of the National Museum of Iceland, sarpur.is. Also see Arthur Middleton Reeves, *The Finding of Wineland the Good: The History of the Icelandic Discovery of America*, Henry Frowde, London, 1890, p. 181, note 66.

NOTES

16 Edward Dwelly (compiler), Derick S. Thomson and Douglas Clyne (eds.), *Appendix to Dwelly's Gaelic-English Dictionary*, Gairm Publications, Glasgow, 1991, p. 59.
17 *Henry IV*, Part I, Act 2, Scene 1.
18 See D. W. Waters, *The Rutters of the Sea: The Sailing Directions of Pierre Garcie*, Yale University Press, New Haven and London, 1967, pp. 40–1 and 208. Also Taylor, *Haven-Finding*, pp. 167–8.
19 I'm quoting from Hardy's *Far from the Madding Crowd*.
20 The words 'month' and 'moon' are related. A month was originally the period of one cycle of the lunar phases, reckoned to be either twenty-nine or thirty days.
21 For a careful analysis, see The Renaissance Mathematicus blog, 'Johannes Kepler's Somnium and Katharina Kepler's Trial for Witchcraft: The emergence of a myth', 30 July 2014, https://thonyc.wordpress.com/2014/07/30/johannes-keplers-somnium-and-katharina-keplers-trial-for-witchcraft-the-emergence-of-a-myth/
22 For examples of what Hardy's villagers could mean, see note 4 to the Introduction.
23 My sense of the pattern is much improved after reading the chapters on the Moon in Thomas Hockey, *How We See the Sky: A Naked-Eye Tour of Day & Night*, University of Chicago Press, Chicago and London, 2011.
24 Bede gives a formula for working out how many hours the Moon will shine each night, with a proof using the full-Moon correlation (*Bede: The Reckoning of Time*, Faith Wallis (trans.), Liverpool University Press, 2004, pp. 73–4). This example was kindly shared with me by Anne Lawrence-Mathers, along with two techniques for knowing the 'houre of the night, by the Moone' in Leonard Digges, *A Prognostication of right good effect ... to judge the wether for ever*, 1555, London; R. T. Gunther (ed.), Old Ashmolean Reprints III, c. 1927, Oxford, p. 63. Local priests, she explained, would have been expected to use these kinds of methods.
25 In the opening chapter of *Far from the Madding Crowd*, Gabriel checked if his watch had slipped, in part, by 'constant comparisons with and observations of the sun and stars'.

26 The date when the Pleiades now reach their highest point (culminate) at midnight is 21 November. See Bruce McClure, 'Halloween is an astronomy holiday', *EarthSky*, 31 October 2025, https://earthsky.org/astronomy-essentials/halloween-derived-from-ancient-celtic-cross-quarter-day/
27 See, for example, the system for telling time by the stars, including by (what we call) the Pleiades and Orion's Belt, used by the Zinacantán community in south-eastern Mexico, in Evon Z. Vogt, 'Zinacanteco Astronomy', *Mexicon*, Vol. 19, No. 6, December 1997, p. 114.
28 Vaiškūnas, 'Celestial Bodies', p. 177.
29 Petriashin, 'Rural timing', p. 872.
30 Vaiškūnas, 'Celestial Bodies', p. 177.
31 This is Henry Thornton Warton's 1887 translation of the 'Midnight Poem', broken into separate lines.
32 There is much debate about the interpretation of the poem. For its expression of time by the events of the sky, I have followed the analysis given by I. S. Herschberg and J. E Mebius in their article, 'ΔΕΔΥΚΕ men a ΣΕΛΑΝΝΑ', *Mnemosyne*, Vol. 43, No. 1/2, Brill, 1990, pp. 150–1. They conclude that the poem describes a night 'just before spring or in very early spring'.
33 Duncan Garrow and Neil Wilkin, *The World of Stonehenge*, British Museum, London, 2022, p. 127.
34 Petriashin, 'Rural timing', p. 867.
35 Ibid., pp. 869 and 872.
36 Vaiškūnas, 'Celestial Bodies', pp. 179 and 180.
37 If we were standing in the centre of a village, each star would appear to rise from or sink behind the same roof or tree every night. Put another way, the rising and setting points of the fixed stars do not move around the horizon, unlike those of the Sun, Moon and planets.
38 Quoted in Seb Falk, *The Light Ages: A Medieval Journey of Discovery*, Penguin, 2021, p. 46.
39 The manual adds that the juniper bush is 'on the path to the well' (ibid., pp. 46–7).
40 See Bonnie Blackburn and Leofranc Holford-Strevens, *The Oxford Companion to the Year*, Oxford University Press, 1999, pp. 595–6.

NOTES

41 After watching Sarah Symons's lecture, 'It's Sirius O'Clock: Astronomical Timekeeping in Ancient Egypt', Royal Astronomical Society of Canada, 10 April 2019, https://www.youtube.com/watch?v=Ak07dI-TALU

42 For an introduction to Sarah Symons's research on this topic, watch her lecture (ibid.), and read her article with Robert Cockcroft, 'Ancient Egyptians measured the first hour, and changed how we related to time', *The Conversation*, 3 July 2023, https://theconversation.com/ancient-egyptians-measured-the-first-hour-and-changed-how-we-related-to-time-203659

43 Specifically, the beginning, middle and end of the night. See O. Neugebauer and R. A. Parker, *Egyptian Astronomical Texts*, Vol. 3, Brown University Press, Providence, RI, 1969, pp. 49–52 and Plate 24.

44 For example, see Anthony Turner, James Nye and Jonathan Betts (eds), *A General History of Horology*, Oxford University Press, 2022, p. xiv, and Jérôme Bonnin, 'Time Measurement in Antiquity', ibid., p. 1.

45 When we spoke in 2025, Sarah explained that in her view, it was most likely not the case, as is often assumed, that the Egyptians first established an hour unit by water clock and then used that to time the stars. Rather, the water clock was a proxy, a timer, which stood in for the *wnwt* when the stars were absent.

46 The ancient Egyptians also measured equal hours, though very rarely. This is shown, for instance, by the ratio tables for day and night in a papyrus published by Abd el-Mohsen Bakir as *The Cairo Calendar No. 86637*, Antiquities Department of Egypt, 1966.

47 At certain times of year, the brightest star Sirius may be seen in daylight.

48 This road runs north to south. I call it Noon Street, for reasons explained on pp. 126-7.

EPILOGUE

1 See Hazel Metherell, 'When is a Dandelion not a Dandelion? (A beginner's guide to yellow composites)', *BSBI News*, No.

THE FULLNESS OF TIME

147, Botanical Society of Britain & Ireland (BSBI.org), April 2021, pp. 37–9.

ACKNOWLEDGEMENTS

1. Jay Griffiths, *Pip Pip: A Sideways Look at Time*, Flamingo, 1999.
2. Paul Glennie and Nigel Thrift, *Shaping the Day: A History of Timekeeping in England and Wales 1300–1800*, Oxford University Press, 2009.

Index

Ahmed, Imad, 158–9
Aldebaran, 167, 184, 187, 195
Allen, Jmeel, 35, 55, 151
almanacs, 183
Arnaldi, Mario, 88–9, 100–2, 104, 107–8
astrolabe, 74
astrological moondial, 40
Aubry, Sylvain, 52–5
Auchindrain, 80
Avebury, 135
axis mundi, 134

Baker Street Irregular Astronomers, 168–70, 179, 182, 188, 190, 193–7, 190
ballads, 60
Barnestone, William, 97
bats, 6, 8, 25–9, 30, 35, 199
　Daubenton's bats, 26, 28
Battersea Power Station Underground Station, 152
bearded tits, 17
Bech, Simon, 117
Bechbretha, 17
Beckenham, Peter, 21
Bede, 160, 183
bees, 17, 29–30, 34, 50–1

Bell, Steve, 139–40, 144–7, 149, 153, 155, 160–8
Belloc, Hilaire, 97
Belt of Venus, 146
Bern Botanical Garden, 52–5
Betelgeuse, 167
Big Dipper, *see* Ursa Major
bird migration, 22–3, 29
Bishopstone church 94–9
bitterns, 24
Björnsson, Árni, 126
blackbirds, 8–9, 11–13, 21, 23, 154, 193
blackcaps, 8, 12, 21–2, 25
bladder campion (*Silene vulgaris*), 47
'blue hour', 154–5
blue tits, 11
Bourne, Henry, *Antiquities of the Common People*, 14
Bräker, Ulrich, 19
Breton, Nicholas, 164
British Summer Time, 176
British Sundial Society, 85, 99
Brú na Bóinne, 129
Buddhist tradition, 157
Bustarfell, 124
butterflies, 9, 34, 200
Byzantines, 192

caddisflies, 26
California poppies (*Eschscholzia californica*), 56
Cambridge Castle, 159
Cambridge University Botanic Garden, 46–8, 52
Campbell, John Lorne, 78
canonical hours, 83–4
Canterbury Pendant, 100–1
Carpathian Mountains, 18
Cassiopeia, 180
cat's-ears (*Hypocheris radicata*), 53, 199
cats' eyes, 18–19
Cerea cactus (*Selenicereus grandiflorus*), 45
chaffinches, 11–12, 21–2
Champion, Matthew, 98
Charles's Wain, *see* Ursa Major
Chaucer, Geoffrey, 15, 33, 73, 103, 143–4
Chichester Cathedral, 111
chicory (*Cichorium intybus*), 8, 34–5, 42, 53–5, 151
China, ancient, 135
church clocks, 109
climate change, 8
coal tits, 12
cockcrow, 1, 14–15, 148
cockerels, 11, 14–17, 20, 24, 29, 158
see also roosters
cockshut, 18
common daisies (*Bellis perennis*), 33–4, 36
common rockrose (*Helianthemum nummularium*), 53
coots, 11
Corfidi, Stephen, 154
Cotton, Charles, 15–17, 24, 102
courgette flowers, 36, 53
Court Lodge Farm, 11–14, 16–17, 30
Covid-19, 4, 23, 76–7
cow parsley, 34

crows, 21
cuckoo clocks, 13
curlew, 9, 67

da Cunha, Alexandre, *Sunset, Sunrise, Sunset*, 152–3
Dales knitters, 68–71
dandelions (*Taraxacum* species), 33–5, 37–9, 199
Darwin, Erasmus, *The Loves of Plants*, 44–6
dawn chorus, 2, 6, 8, 12–13, 22, 179
'day' (the word), 164
daybreak, 50, 76, 165
dæg-wōma ('day-noise'), 11–12
daylight-saving time, 122
daymarks, 113–14, 119–21, 123–7, 137–8, 178, 189, 195
de Candolle, Augustin Pyramus, 49
de Mairan, Jean-Jacques d'Ortous, 49
Dentdale, 68, 70–1
Dial Hill/Dial Pasture, 105
Digby, Sir Kenelm, 72
'dog days', 190
Dragon Hill, 130–2
ducks, 3, 11, 95
Dunlop, Frances, 61–4, 66–7, 79–81
dunnocks, 11

Earth's axis, 96
Earth's poles, 143
echolocation, 26
Egeria, 165
egrets, 24
Egyptians, ancient, 190–3, 196
Eid, 159
eleven-o'clock-lady (Star-of-Bethlehem) (*Ornithogalum umbellatum*), 53
Emms, Eric, 169–70, 174, 185, 188, 194
Equator, 83

INDEX

equinoxes, 83, 182
Eurasian plate, 115
evening primroses (*Oenothera*), 32–3
evensong, 22, 25
eyes, sensitivity to light and colour, 162–3
eyktir, 119–21, 124

Falk, Seb, 189
Feltham Marshalling Yards, 24
floral clocks, 42–6, 52–5
Florio, John, *A Worlde of Words*, 18
Forby, Robert, *The Vocabulary of East Anglia*, 156
'forlongwey', 73
four o'clocks (*Mirabilis jalapa*), 36
French Revolution, 46

Garcie, Pierre, *Le grant routtier*, 181
Gazania, 53
Gentleman's Magazine, 127
Gerard, John, 38–9, 41
giant waterlily (*Victoria cruziana*), 51–2
Gioia, Ted, 77
gloaming, 139–40
Glover, Beverley, 51–2, 54
goat's beard (Jack-go-to-bed-at-noon) (*Tragopogon pratensis*), 27, 35–6, 38, 53–4, 56, 179
goats, 19, 122
Gockerell, Nina, 72, 125
  'Telling the Time Without a Clock', 18–19
gökotta, 21–2
goldfinches, 11
Grágás laws, 116–18
Grahame, Kenneth, *The Wind in the Willows*, 4
Great Bear, *see* Ursa Major
great tits, 21
Greece, ancient, 135, 192
Greenwich Mean Time, 6, 126, 158

Greenwich Meridian, 142
Grigson, Geoffrey, 105
gulls, 22, 57, 60

hádegi ('high-day'), 121
hairstyles, horticultural, 44
Halloween, 186
Hammersmith Bridge, 126
han-crēd (cockcrow), 14
Hardy, Thomas, *A Pair of Blue Eyes*, 7, 20
  Far from the Madding Crowd, 7, 171–6, 178–80, 183–4, 194
  The Woodlanders, 147
Harland, John, *Reeth Bartle Fair*, 71
Hartley, Marie, 69–70
hawkweed (*Hieracium umbellatum*), 42
haymaking, 122
Hazlitt, William, 7
Hemans, Felicia, 'The Dial of Flowers', 46
herons, 24
Highland Clearances, 80
Hill, Thomas, 37
HM Nautical Almanac Office, 141, 144, 156, 158
honeysuckle (*Lonicera periclymenum*), 29, 31, 56
hummingbirds, 51
Hyades, 184

Iceland, 113–26, 136–8, 188–9
  change to 'English time', 125–6
  see also daymarks
ice-plant (*Mesembryanthemum crystallinum*), 42
Ingilby, Joan, 69
Islamic calendar, 157
Islamic timekeeping, 143, 157–60
Isle of Barra, 58, 66
Isle of Eriskay, 58
Isle of South Uist, 65, 66, 78

jasmine (*Jasminum officinale*), 31, 56
Joannou, Nick, 170, 183
Johnston, Annie, 58, 63, 67
Jones, Ben, 85, 89–90, 99
Jupiter, 160, 169

Kålund, Kristian, 117
Kentish drovers, 103
Kepler, Johannes, *Somnium*, 183
King's Mirror (*Konungs skuggsjá*), 136
Kircher, Athanasius, 40–1
Kirkdale, St Gregory's Minster, 84–5, 88
Kristjánsdottir, Sigrún, 122
kveldvaka ('evening wake'), 177

lacemakers, 69
lacewings, 26
lágnætti ('low-night'), 121
Lalande, Jérôme, 156–7
lapwing, 67
larks, 11–13, 15–16, 22, 24, 110
Lárusdóttir, Birna, 124, 126–7, 137
Lawrence-Mathers, Anne, 184
Le Calendrier des Bergers, 176
light pollution, 29, 140
Linnaeus, Carl, 42–4, 49, 52–4
London Underground, 152
London Wetland Centre, 24
long-tailed tits, 145
Lord's Prayer, 72
Lovett, Edward, 105–7, 109, 111–12
low light visibility, 144
lunar phases, 183
Lynch, Kevin, *What Time Is This Place?*, 151–2

Mackenzie, Fiona J., 63–5, 79
MacRae, Peigi and Màiri, 65, 78
Maghrib, 157
marigolds (*Calendula officinalis*), 33, 37–9, 45, 50
marsh harriers, 17

Marvell, Andrew, 41, 46
Maurer, Konrad, 121
mayflies, 26
mealtimes, 76
Met Office, 2
Methúsalemsson, Methúsalem, 124
Mid-Atlantic Ridge, 115
Midday Peaks, 125
midges, 25–6
midsummer, 7, 83, 103, 115–17, 121, 128–9, 132–6, 138
midwinter, 83, 128–32, 136, 172, 175, 185
'mileway', 73–5
milk lorries, 109
milking, 16–17, 65
Milky Way, 184
Milton Keynes, 132–6
Minnaert, Marcel, *Light and Colour in the Outdoors*, 147–50, 153–6, 166–7
Miserere Psalm, 72
Mondschein, Ken, 70
moonlight, 12, 45, 51, 141–2, 162
morning glory (*Ipomoea purpurea*), 47, 50, 53
mosquitoes, 26
Mosscrop, Stuart, 135
moths, 13, 26–7, 29, 50–1, 193
mugwort, 34

Nairobi, 175
Nautical Almanac, 141–3, 147, 156
New Crescent Society, 158–9
New Year's Day, 127–8
Newgrange passage tomb, 129–30
Newton, Elliot, 25–8
night-flowering catchfly (*Silene noctiflora*), 53
nightingales, 24
night-scented stock (*Matthiola longipetala*), 56
Nile, river, 190

INDEX

nipplewort (*Lapsana*), 45
noon (the word), 92
Noon Stone, 125
Noon Street, 127–8, 137–8, 193
Noon Sun Hill, 125
Norman Conquest, 85
North American plate, 115
North Star (Polaris), 96–7, 169,
   171–6, 179–81, 194
   Guardian stars, 173–5, 179, 181

Old Kent Road, 103
Orion, 167, 169, 174, 188, 194–6
Orion's Belt, 188
ornithological clock, 12–13
owls, 24, 29, 165–6, 190
   barn owl (*Tyto alba*), 24
Öxará river, 116
ox-eye daisies, 27

Parisian fullers, 165
Parisian leatherworkers, 157
Pevensey Marshes Nature Reserve, 17
photosynthesis, 47–8
Pleiades (Seven Sisters), 178, 184–9,
   194–5, 197
Pliny the Elder, 37
Plough, the, *see* Ursa Major
Polaris, *see* North Star
Pollard, Josh, 129–31
Poller, Frank, 90–1
pollinators, 50–2, 54–6
Proctor, Richard, 172–5, 196
Purkinje Effect, 163
purple shamrock (*Oxalis
   triangularis*), 35, 55

quail, 12
querning songs, 66–7

Ramadan, 159
Reykjavík, 114, 121, 123, 175
Ridgeway, 82, 99–101, 105–6, 109–11

robins, 2, 11, 13, 21–2, 200
Rökkurró, 138
Roman Empire, 192
Roman hours, 83, 88, 92
rooks, 11, 109
roosters, 14, 18, 50, 171
   see also cockerels
Royal Greenwich Observatory, 126
Rural Cyclopedia, The, 38

sacred timings, 144
St Dunstan, 101
San Francisco, 135
Sappho, 186
scarab beetles, 52
scarlet pimpernel (*Anagallis
   arvensis*), 37, 42
Sedgwick, Adam, 70
sensitive plant (*Mimosa pudica*), 49
Sesto Sundial, 125
Sgioba Luaidh Inbhirchluaidh,
   61–2, 79–81
Shakespeare, William, 15, 105, 180
Shaw, Margaret Fay, 63–4, 78–9
sheep, 3, 17–21, 57, 59, 61, 67–8, 70,
   105–6, 122, 171
   eye-pupils, 18–21
   lambing, 18
shepherds, 5–6, 9, 15, 18–19
   Italian, 104
   Lithuanian, 103–4
   Pennine dales communities, 68, 70
   'red sky at night', 154
   'shepherd's clock', 37
   South Downs, 105–9, 111–12
   and star clocks, 171, 176, 178, 180,
      184
Sigurðsson, Gísli, 114–18, 123–4, 138,
   164
Silbury, 135
Silva, Fabio, 178
Simmer Dim, 143
Sirius (Dog Star), 190, 192–6

snipe, 67
Society Islands, 165
solstices, 127–8, 133, 135, 172, 175–6, 182, 200
sparrows, 12, 30, 199
starlings, 23–4
stars, 1–2, 4, 7, 23, 74, 138, 141–3, 145, 166–7, 169, 171–85, 187–96, 199–200
  circumpolar stars, 174, 176, 180
  star tables, 143, 190–1, 196
Steinólfsdóttir, Halla, 122
Stereochron Island, 21, 23
Stonehenge, 128, 132
streetlamps, 151–2
Styria, 72
'sun marks', 124–5
sundials, 5, 7, 40, 82–112, 121, 125, 127, 134, 137, 199
  'horologium viatorum', 102
  Mass dials, 89, 91
  scratch dials, 82–3, 85–91, 94–6, 98, 100, 103
  umbra sumus motto, 98
sunflowers (*Helianthus annuus*), 31, 38–41, 49–50
sun-spurge (*Euphorbia helioscopia*), 39
Symons, Sarah, 191–3, 195

tattoos, 131
Taurus, 167, 184
Taylorism, 77
Tennyson, Alfred, 46
thrushes, 9, 22, 25
tobacco plants (*Nicotiana*), 50–1
toothache, 73
Tragopogon porrifolius, 47
trumpet gentian (*Gentiana acaulis*), 53
twilight, 7, 22, 24–5, 48, 138–47, 149, 152, 155–7, 161–2, 164–9, 193, 199
  'between dog and wolf', 18, 165
  civil twilight, 142–3, 149, 155–6

cwyld-seten, 161
'dark hour', 156–7
nautical twilight, 142–3, 161, 166
'purple light', 149–50

Ukraine, 160
urination, 73
Ursa Major, 169, 171–2, 174–9, 189, 191, 193–4, 196
US Naval Observatory, 141

Venus, 160
Vilhjálmsson, Thorsteinn, 121, 134

Waddell, Mark, 41
Walker, George, 68
Walpole, Horace, 45
waterlily (*Nymphaea alba*), 42, 45, 51–2
Waugh, Albert, 127
waulking songs, 57–64, 78–81
Wayland's Smithy, 99, 131
Webb, Alex, 46–8, 52, 54
Wensleydale, 68
White Horse of Uffington, 110–12, 114, 130–2
'witch marks', 94
wnwt, 191–2
wolves, 164
  'between dog and wolf', 18, 165
wood pigeons, 21
wood sorrel (*Oxalis corniculata*), 36
Woolstone church, 82–3, 85–95, 99–100, 111–12, 130–1, 134
wrens, 21, 23
Wycherley, William, *The Plain Dealer*, 73

yellowhammers, 19
York Minster masons, 74–6

zodiac, 185
Zugunruhe, 22

A Note on the Author

Cathy Haynes is a curator, writer, artist and educator who has been developing a creative practice on aspects of time for more than two decades. She has been Timekeeper in Residence at the Petrie Museum of Egyptian Archaeology, artist in residence in Victoria Park for the Chisenhale Gallery, Curator for Art on the Underground (Transport for London) and a founding faculty member at The School of Life. She has contributed to the *Guardian*, *The Human Zoo* on BBC Radio 4 and *The Monocle Weekly*. She lives in London.

A Note on the Type

The text of this book is set in Adobe Garamond. It is one of several versions of Garamond based on the designs of Claude Garamond. It is thought that Garamond based his font on Bembo, cut in 1495 by Francesco Griffo in collaboration with the Italian printer Aldus Manutius. Garamond types were first used in books printed in Paris around 1532. Many of the present-day versions of this type are based on the *Typi Academiae* of Jean Jannon cut in Sedan in 1615.

Claude Garamond was born in Paris in 1480. He learned how to cut type from his father and by the age of fifteen he was able to fashion steel punches the size of a pica with great precision. At the age of sixty he was commissioned by King Francis I to design a Greek alphabet, and for this he was given the honourable title of royal type founder. He died in 1561.